Essential Maths Skills
for AS/A-level
Psychology

Molly Marshall

PHILIP ALLAN FOR
 HODDER
EDUCATION
AN HACHETTE UK COMPANY

Philip Allan, an imprint of Hodder Education, an Hachette UK company, Blenheim Court, George Street, Banbury, Oxfordshire OX16 5BH

Orders

Bookpoint Ltd, 130 Park Drive, Milton Park, Abingdon, Oxfordshire OX14 4SE

tel: 01235 827827

fax: 01235 400401

e-mail: education@bookpoint.co.uk

Lines are open 9.00 a.m.–5.00 p.m., Monday to Saturday, with a 24-hour message answering service. You can also order through the Hodder Education website: www.hoddereducation.co.uk

ISBN 978-1-4718-6353-0

First printed 2016
Impression number 5 4 3 2 1
Year 2020 2019 2018 2017 2016

Typeset in India

Cover illustration: Barking Dog Art

Printed in Spain

Hachette UK's policy is to use papers that are natural, renewable and recyclable products and made from wood grown in sustainable forests. The logging and manufacturing processes are expected to conform to the environmental regulations of the country of origin.

Contents

The listed content is assessed by the awarding bodies AQA, OCR, Edexcel and WJEC/Eduqas at AS and A-level.

The content listed in bold is only specified to be assessed at AS level by OCR. For all other boards, these skills are assessed at A-level only.

3 Algebra

4 Graphs

Exam-style questions

Appendix 1

Appendix 2

Introduction

In order to be able to develop skills, knowledge and understanding in psychology, students need to have acquired competence in a range of areas in mathematics. Both A-level and AS level examinations in psychology require the use of maths and, because at least 10% of the marks in all awarding body assessments for psychology requires essential maths skills to at least the standard of higher tier GCSE mathematics, it is important that students acquire competence in these skills.

This book is aimed at students who are struggling with mathematics, assumes little prior knowledge and is a guide to the maths skills students should understand and be able to demonstrate in A-level and AS level examinations. The guide does not include any stretch questions because the aim is to help students to be able to answer the sort of mathematics questions that might arise in a psychology exam. The guide is intended as a revision aid, not a textbook, and it focuses on developing the understanding of maths skills so that students can read and understand psychological research studies and design, carry out and analyse the data from their own independent research projects.

The guide includes every mathematical topic listed as a requirement (by AQA, OCR, Edexcel and WJEC/Eduqas), worked examples showing how the maths skills can be applied in psychological research, guided and practice questions and exam-style questions showing how the maths skills content could be examined and how answers might be assessed.

The four topics covered in the guide are arithmetical and numerical computation, handling data, algebra and graphs. For each of these the following are provided:

- explanations and examples of the maths skills in the context of A-level and AS level psychology
- worked examples showing how the maths skills can be applied in the context of psychology
- guided and non-guided practice questions
- exam-style questions, with the appropriate breakdown of marks

Full worked solutions to the guided and practice questions and exam-style questions can be found online at www.hoddereducation.co.uk/essentialmathsanswers.

1 Arithmetic and numerical computation

Expressions in decimal and standard form

Decimal form

A decimal is any number in our base-10 number system. A base number is the basis of a place value number system, in which successive powers of the base number are used for each column. The decimal system uses 10 as its base number so it is called a base-10 system. The decimal point is used to separate the ones place from the tenths place in decimals. (It is also used to separate pounds from pence in money.) As we move to the right of the decimal point, each number place is divided by 10 (see Table 1.1).

Table 1.1 Place value and decimals

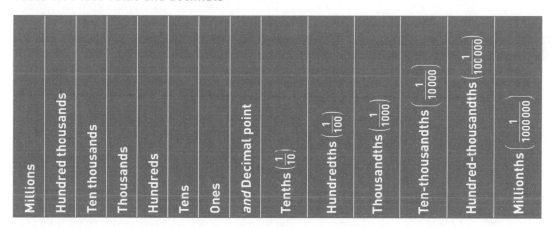

For example, thousands divided by 10 gives you hundreds. This is also true for digits to the right of the decimal point. For example, tenths divided by 10 gives you hundredths. Thus, we read the decimal 59.46 as 'fifty-nine and forty-six hundredths'. Note that usually we read the decimal point as 'point' so, 59.46 would be read as 'fifty-nine point four six'.

Example

Table 1.2 shows that each phrase can be written as a fraction and as a decimal.

Table 1.2

Phrase	Fraction	Decimal
Six tenths	$\frac{6}{10}$	0.6
Seven hundredths	$\frac{7}{100}$	0.07
Fifty-two hundredths	$\frac{52}{100}$	0.52
Three hundred and eighty-seven thousandths	$\frac{387}{1000}$	0.387

Decimals are used in calculations that require more precision than whole numbers can provide. A good example of this is money. Three and one quarter pounds is an amount of money and we use decimals to write this amount as £3.25.

A decimal may have both a whole-number part and a fractional part. The whole-number part of a decimal is those digits to the left of the decimal point. The fractional part of a decimal is represented by the digits to the right of the decimal point.

A Worked example

Complete Table 1.3 with the whole-number part and fractional part of each decimal.

Remember the whole-number part is to the left of the decimal point and the fractional part is to the right of the decimal point. Table 1.4 shows the answers.

Table 1.3

Decimal	Whole-number part	Fractional part
4.45		
6.272		
28.04		
0.565		
256.8		

Table 1.4

Decimal	Whole-number part	Fractional part
4.45	4	45 hundredths
6.272	6	272 thousandths
28.04	28	04 hundredths
0.565	0	565 thousandths
256.8	256	8 tenths

Look at these decimals in the place-value chart (Table 1.5).

Table 1.5 Place value and decimals

Millions	Hundred thousands	Ten thousands	Thousands	Hundreds	Tens	Ones	and Decimal point	Tenths ($\frac{1}{10}$)	Hundredths ($\frac{1}{100}$)	Thousandths ($\frac{1}{1000}$)	Ten-thousandths ($\frac{1}{10000}$)	Hundred-thousandths ($\frac{1}{100000}$)	Millionths ($\frac{1}{1000000}$)
						4	.	4	5				
						6	.	2	7	2			
					2	8	.	0	4				
						0	.	5	6	5			
				2	5	6	.	8					

Note that 0.565 has the same value as .565. The zero in the ones place helps us remember that 0.565 is a number less than one.

B Guided question

Copy out the workings and complete the answers on a separate piece of paper.

1 Write each decimal from Table 1.6 into the place value chart (Table 1.7).
The first one has been done for you.

Table 1.6

Phrase	Decimal
Forty-three hundredths	0.43
Six tenths	0.6
Sixty and five hundredths	60.05
Thirty and sixty-six hundredths	30.66
Twenty two and seventy-two thousandths	22.072

Table 1.7 Place value and decimals

Millions	Hundred thousands	Ten thousands	Thousands	Hundreds	Tens	Ones	*and* Decimal point	Tenths $\left(\frac{1}{10}\right)$	Hundredths $\left(\frac{1}{100}\right)$	Thousandths $\left(\frac{1}{1000}\right)$	Ten-thousandths $\left(\frac{1}{10\,000}\right)$	Hundred-thousandths $\left(\frac{1}{100\,000}\right)$	Millionths $\left(\frac{1}{1\,000\,000}\right)$
						0	.	4	3				

C Practice questions

2 Which of the following is equal to $\frac{7}{100}$?

A 7

B 0.7

C 0.07

D 0.007

3 Which of the following is equal to $\frac{550}{1000}$?

A 0.550

B 5.500

C 0.0550

D 0.00550

4 Which decimal represents six hundred and sixty-six and six tenths?
   A 0.6666
   B 666.6
   C 66.66
   D 666.06

5 Which of the following is equal to seventy and sixty-nine thousandths?
   A 70.069
   B 70.0069
   C 70.690
   D 70.00069

Standard form

In psychology you probably won't have to write down very large or very small numbers but when reading psychological research you need to be able to look at a table of results and draw conclusions from the quantitative data. Also, you need to be able to construct (and interpret) charts and diagrams from quantitative data.

Standard form is a way of writing very large or very small numbers easily. They are written as a calculation which multiplies a number between 1 and 10 by the appropriate power of 10 to make the very large or very small number.

For example, $10^3 = 1000$, so $4 \times 10^3 = 4000$. So 4000 can be written as 4×10^3. It is useful to be able to write very large numbers in standard form.

Small numbers can also be written in standard form. However, instead of the index being positive (in the above example, 10^3, the index is $^{+3}$), it will be negative (as in 10^{-2}). (The index is the little number that is raised up.)

When writing a number in standard form:
- first you write a number between 1 and 10
- then you write $\times 10^{\text{(to the power of a number)}}$

Examples

- 82 800 000 000 000 written in standard form is 8.28×10^{13}

8.28 multiplied by 10^{13} results in the number 82 800 000 000 000.
It's 10^{13} because 8.28 has to be multiplied by ten 13 times to get the number to be 8.28.

- 0.0000015 written in standard form is 1.5×10^{-6}

1.5 multiplied by 10^{-6} results in the number 0.0000015. This can also be thought of as dividing by 10^6.
It's 10^{-6} because 1.5 has to be divided by ten 6 times to get the number to be 1.5.

- 150 written in standard form is 1.5×10^2.
- 2500 written in standard form is 2.5×10^3.
- 150 + 350 written in standard form is $(1.5 \times 10^2) + (3.5 \times 10^2)$.
- 1000 + 505 written in standard form is $(1 \times 10^3) + (5.05 \times 10^2)$.

On a calculator, you *usually* enter a number in standard form as follows:

- Type in the first number (the one between 1 and 10).
- Press EXP.
- Type in the power to which the 10 is raised.

Practise entering numbers in standard form on your calculator.

> **REMEMBER**
>
> In multiplication calculations involving indices you ADD the indices. For example:
>
> $p^3 \times p^7 = p^{10}$
>
> In division calculations involving indices you SUBTRACT the indices. For example:
>
> $p^5 \div p^3 = p^2$

Ⓐ Worked examples

a Calculate $4x^3 \times 3x^2$.

The numbers in front of the variables follow the usual rules of multiplication and division, but index numbers follow the rules of indices.

Step 1: multiply the first part of each number.

$4 \times 3 = 12$

Step 2: add the indices.

$3 + 2 = 5$

Thus $4x^3 \times 3x^2 = 12x^5$

b The number a written in standard form is 6×10^5. The number b written in standard form is 7×10^{-2}.

Calculate $a \times b$ and write the answer in standard form.

You can use your calculator to do this.

Step 1: multiply the first part of each number.

$6 \times 7 = 42$

Step 2: multiply the second part of each number.

$10^5 \times 10^{-2} = 10^3$

Step 3: combine the two answers.

42×10^3

Step 4: convert the answer to standard form. The first part of the answer (42) should be a number between 1 and 10.

$42 \times 10^3 = 4.2 \times 10^4$

B Guided question

Copy out the workings and complete the answers on a separate piece of paper.

1 **Table 1.8 shows a list of numbers to be written in standard form.**

    Complete the table and then use your calculator to check these are correct. The first two have been done for you.

Table 1.8

Decimal form	Standard form	What calculation needs to be done
4200	4.2×10^3	Multiply by 1000
0.00085	8.5×10^{-4}	Divide by 10 000
15.5		
0.00888		
5 500 000 000		

C Practice question

2 Does noise affect concentration? In a research project, three groups of students revised a psychological study of obedience for 10 minutes. Group A revised in a classroom in silence, Group B revised in a noisy sixth form common room and Group C revised in a busy canteen. Immediately after revising all three groups were asked 20 questions on the study. Each correct answer scored one point.

The average scores for each group are shown in Table 1.9.

Table 1.9 Results of the project

	Group A	Group B	Group C
Conditions	Silence	Noisy common room	Busy canteen
Average score	16.3	16.05	14.68

a Which group scored the highest?

b Which group scored the lowest?

c Write the results in standard form:

Group A =

Group B =

Group C =

Ratios, fractions and percentages

Ratios

In mathematics, a ratio is a relationship between two numbers of the same kind. For example, objects, persons, students, fruits, etc. The relationship is expressed as a to b or a:b, indicating how many times the first number (a) contains the second number (b).

In simple terms, a ratio represents, for every amount of one thing, how much there is of another thing. Ratios are always written in the simplest form.

Example

If you have 8 apples and 2 bananas:

Figure 1.1

the ratio of apples to bananas is 8:2 but because 8 can be divided by 2, and 2 can be divided by 2, the ratio can be simplified to 4:1.

$8 \div 2 = 4$

$2 \div 2 = 1$

Figure 1.2

The ratio of bananas to apples is 2:8, which can be simplified to 1:4.

What is the ratio of apples to fruits?

There are 10 fruits (8 apples + 2 bananas) so the ratio of apples to the total amount of fruit is 8:10, which can be simplified to 4:5.

The 4:5 ratio can be converted to a fraction of $\frac{4}{5}$ to represent how much of the fruit is apples.

Example

Figure 1.3

If you have 10 toy cars and 8 toy lorries, the ratio of cars to lorries is 10:8. The numbers 10 and 8 can both be divided by 2 (2 is the highest common factor) so the ratio can be simplified to 5:4.

The ratio of 5:4 cannot be simplified because there is no number that will divide into 5 and also divide into 4 (except 1).

Worked example

A psychologist handed out questionnaires to an opportunity sample of 60 men and 40 women.

What is the ratio of male to female participants?

A 6:4

B 4:6

C 3:2

D 2:3

- Both A and C show the ratio of male to female participants, but C is the better answer because 6:4 can be simplified (both 'sides' can be divided by 2) and shown as 3:2.
- For every three male participants there were two female participants, so the sample was a gender-biased sample.

B Guided question

Copy out the workings and complete the answers on a separate piece of paper.

1 Table 1.10 shows four situations in which a ratio can be calculated, which ratio is to be calculated and how to calculate the ratio.

   Look at the two rows that have been completed for you and then calculate the ratios for the other two. Give your answers in simplest form.

Table 1.10

Situation	Ratio required	Answer
I have 20 bottles of red wine and 4 bottles of white wine.	Red wine to white wine	20:4 which can be simplified because 20 and 4 are both divisible by 4. The ratio is 5:1.
In our road people own 24 cats and 8 dogs.	Dogs to cats	8:24 which can be simplified because 8 and 24 are both divisible by 8. The ratio is 1:3.
Henry planted 4 tomato plants and 10 strawberry plants.	Strawberry plants to the total number of plants	
Tom received 120 votes and Harry received 200 votes.	Tom's votes to the total number of votes	

C Practice question

2 A psychologist was interested to find out how people used their mobile phones in public places. For three weekends she and two other researchers carried out an observational study in a large shopping centre. Each time a passer-by was seen using a mobile phone an entry was made on a tally chart (see the example in Table 1.11).

Table 1.11 Sample tally chart showing mobile phone use

Male/female	Selfie	Photo	Texting	Talking	GPS	Other
M	X	X				
M			X		X	
F		X		X		
F			X	X		
F				X		X

The data showed that:

- 350 women and 140 men were observed using a mobile phone.
- 45 women and 18 men used their phone to take a 'selfie'.
- 240 women and 80 men used their phone to talk.

a   In total, how many people were observed?

b   What was the ratio (in simplest form) of female to male participants?

   A   5:2　　　　　　　　　　C   3.5:1.4

   B   35:14　　　　　　　　　D   3:1

c   What was the ratio (in simplest form) of male to female participants using their phone to take a selfie?

   A   5:2　　　　　　　　　　C   18:45

   B   2:5　　　　　　　　　　D   18:63

d   What was the ratio (in simplest form) of female to male participants using their phone to talk?

   A   24:8　　　　　　　　　　C   18:24

   B   3:1　　　　　　　　　　D   6:4

Fractions

You need to be able to convert ratios to fractions. Fractions are simple! If you eat $\frac{1}{4}$ of a cake you have eaten 1 of 4 parts. If you eat $\frac{1}{2}$ of a bar of chocolate you have eaten 1 of 2 parts. If you share your cake equally with 7 friends, each of you has eaten $\frac{1}{8}$ or 1 of 8 parts.

Converting ratio to fractions

Examples

1   An observer recorded that the ratio of silver cars to red cars in a car park was 3:1 (3 silver cars for 1 red car).

Since 3 'parts' + 1 'part' = 4 'parts'

3 of the 4 parts or $\frac{3}{4}$ of the cars were silver.

1 of the 4 parts or $\frac{1}{4}$ of the cars were red.

2   A supermarket worker counts the tins of soup on the shelf.

20 tins of soup are tomato flavour and 15 tins of soup are chicken flavour.

The ratio of tomato soup to chicken soup is 20:15 which can be simplified to 4:3, so in total there are 7 'parts'.

So, $\frac{4}{7}$ of the tins are tomato soup and $\frac{3}{7}$ of the tins are chicken soup.

3 A gardener kept a record of the number of days it rained in April.

He noted that it rained on 10 days and was dry on 20 days.

So, the ratio of wet days to dry days was 10:20 or simplified as 1:2 so there are 3 'parts'.

As fractions this means $\frac{1}{3}$ were wet days and $\frac{2}{3}$ were dry days.

Ⓐ Worked example

A psychologist handed out questionnaires to an opportunity sample of 80 men and 40 women, so the ratio of male to female participants is 2:1.

What fraction of the sample are male?

What fraction of the sample are female?

Step 1: there are 3 'parts' in total because the ratio of male to female participants is 2:1.

Step 2: converted to a fraction, $\frac{2}{3}$ of the sample are male and $\frac{1}{3}$ of the sample are female.

Ⓑ Guided question

Copy out the workings and complete the answers on a separate piece of paper.

1 100 men completed a psychometric test to measure their personality type. The pie chart shows the results of the test.

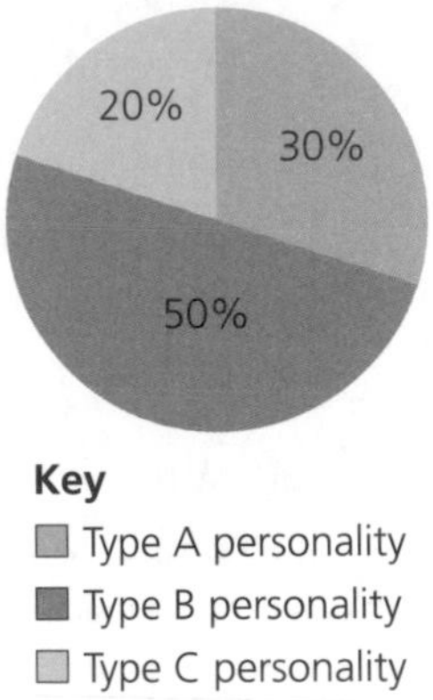

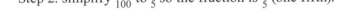

Key

■ Type A personality
■ Type B personality
□ Type C personality

Figure 1.4

a What fraction of the men tested have a type C personality?

Step 1: from the pie chart, identify the percentage of men with a type C personality.

20%, which is $\frac{20}{100}$

Step 2: simplify $\frac{20}{100}$ to $\frac{1}{5}$ so the fraction is $\frac{1}{5}$ (one fifth).

b What fraction of the men tested have a type B personality?

A $\frac{1}{5}$ C $\frac{2}{5}$

B $\frac{3}{10}$ D $\frac{1}{2}$

Step 1: from the pie chart, identify the percentage of men with a type B personality.

Step 2: how can this percentage be written as a fraction in simplest form?

Ⓒ Practice question

2 A psychologist carried out a survey of all his students. The critical question on the questionnaire was:

How interesting do you find psychology?

Very interesting? Interesting? Quite interesting? Not interesting?

The responses were as follows:

Very interesting: 50

Interesting: 40

Quite interesting: 20

Not interesting: 10

a Which of A, B, C or D shows the ratio of 'Very interesting' to 'Not interesting' responses? (Select the answer in the simplest form.)

    A 50:10   C 50:1

    B 5:1   D 10:5

b Which of A, B, C or D shows the ratio of 'Very interesting' to 'Interesting' responses? (Select the answer in the simplest form.)

    A 5:4   C 50:90

    B 50:40   D 4:5

c The total 'parts' are 12 because $5 + 4 + 2 + 1 = 12$. The table below shows the ratio of all the responses.

Table 1.12 Results of the project

Very interesting	Interesting	Quite interesting	Not interesting
5:12	4:12	2:12	1:12

Write as a fraction:

i The number of participants who said that psychology was 'Very interesting'.

ii The number of participants who said that psychology was 'Interesting'.

iii The number of participants who said that psychology was 'Quite interesting'.

Percentages (%)

In psychology, you need to be able to calculate percentages. For example, if you are carrying out observational research you will probably want to know what percentage of the behaviours you observed fell into specific categories. Or if you are doing a survey, handing questionnaires to an opportunity sample, you may want to calculate what percentage of participants were male or female. If you have studied Milgram's research into obedience, you will know that $\frac{26}{40}$ participants administered a 450 V electric shock and this can also be shown as 65%.

What does percent mean? When you say 'percent' you are really saying 'per 100'. One percent (1%) means 1 per 100, 4% means 4 per 100, 50% means 50 per hundred, etc. If a bank offers a rate of 4% per year on a savings bond, it means that for every £100 saved it will pay £4 interest.

Converting fractions to percentages

Some fractions are easy to convert to percentages.

- $\frac{1}{100} = 1$ per hundred and is 1%

- $\frac{1}{10} = 10$ per hundred and is 10%

- $\frac{1}{2} = 50$ per hundred and is 50%

- $\frac{1}{4} = 25$ per hundred and is 25%

Some fractions are less easy to convert. However, there is a method to do the conversion.

(A) Worked examples

a Convert $\frac{4}{5}$ to a percentage.

Step 1: divide the numerator (top) of the fraction by the denominator (bottom).

$4 \div 5 = 0.80$

Step 2: multiply the result of Step 1 by 100.

$0.80 \times 100 = 80$

Step 3: round the answer if needed (usually to two decimal places).

Step 4: add the % sign so the fraction $\frac{4}{5}$ as a percentage is 80%.

b A psychology student decided to carry out a survey and handed questionnaires to 300 students. The critical question was:

How difficult do you find maths?

Very difficult? Difficult? Quite difficult? Not difficult?

The responses were as follows:

	Very difficult	Difficult	Quite difficult	Not difficult
Responses	110	90	60	40

Work out the ratios and percentages for each response.

Step 1: ratios are

	Very difficult	Difficult	Quite difficult	Not difficult
Ratios	11	9	6	4

Step 2: percentages are

	Very difficult	Difficult	Quite difficult	Not difficult
Percentages	$\frac{110}{(110 + 90 + 60 + 40)}$ $\times 100 = 36.7\%$	$\frac{90}{(110 + 90 + 60 + 40)}$ $\times 100 = 30\%$	$\frac{60}{(110 + 90 + 60 + 40)}$ $\times 100 = 20\%$	$\frac{40}{(110 + 90 + 60 + 40)}$ $\times 100 = 13.3\%$

c A sample of 150 people were asked whether they had ever tried to lose weight and 55 people said they had. What percentage of the sample had tried to lose weight?

Step 1: 55 tried to lose weight, divided by the total sample of $150 = 0.36666....$

Step 2: multiply 0.36666 by $100 = 36.66\mathbf{6}$

Step 3: round up to two decimal places so the answer is 36.67%.

d A sample of 200 students were asked whether they were 'good at' mathematics and 45 said yes. What percentage of the sample of students said they were good at mathematics?

Step 1: 45 said yes. Divide 45 by the total sample of 200.

$45 \div 200 = 0.225$

Step 2: multiply 0.225 by 100.

$0.225 \times 100 = 22.5$

Step 3: the percentage who said they were good at mathematics was 22.5%.

B Guided question

Copy out the workings and complete the answers on a separate piece of paper.

1   **85 parents out of a sample of 240 said that their 7-year-old son had his own mobile phone. What percentage of parents said their son had a mobile phone?**

Step 1: identify how many parents said their son had a mobile phone. How many were in the sample?

Step 2: divide the number who said their son had a mobile phone by the number in the sample.

Step 3: multiply the fraction from Step 2 by 100 to find the percentage.

Step 4: give your answer to two decimal places.

C Practice question

2   A psychologist was interested to find out how people use their mobile phones in public places. Researchers carried out an observational study in a large shopping centre. Each time a person was seen using a mobile phone, an entry was made on a tally chart. The data collected is shown in the table below:

Table 1.13 Mobile phone use in a public place

Behaviour	Male	Female	Total
Selfie	18	45	63
Photo	5	21	26
Texting	45	85	130
Talking	82	240	322
GPS	15	5	20
Total	165	396	561

a   In total, how many males were observed?
b   In total, how many females were observed?

For each of the questions below, give your answer to 2 d.p.

c   What percentage of the observed participants were male?
d   What percentage of the observed participants were female?
e   What percentage of the observed participants used their phone for a selfie?
f   What percentage of the observed participants used their phone to text?
g   What percentage of the observed participants used their phone to talk?

Estimating results

Estimating is a useful tool in everyday life. Get in the habit of estimating amounts of money, lengths of time and distances. Being able to estimate is especially useful in psychology, because there is not a lot of point carrying out complicated statistical analysis if there is little or no difference between the two (or more) sets of scores. Being able to have a 'quick look' at a set of data and estimate results accurately can save a lot of time.

You can think of estimating as 'rounding down' or 'rounding up' and you can use significant figures to get an approximate answer to a problem. For example, you buy a birthday present costing £4.99 and then you spend £2.99 on coffee and cake. By rounding up to the nearest whole pound it's easy to estimate that you have spent nearly £8.

(£4.99 rounded up to £5) + (£2.99 rounded up to £3) = £8

You can round up (or down) all the numbers in a mathematics problem to the nearest whole number to make 'easier' numbers. It is usually possible to do this in your head!

A Worked examples

a Estimate the result of 8.9 divided by 3.07.

$$\frac{8.9}{3.07}$$

Step 1: round the numerator (8.9) to the nearest whole number.

Numerator = 9

Step 2: round the denominator (3.07) to the nearest whole number.

Denominator = 3

Step 3: divide the numerator by the denominator.

$$\frac{9}{3} = 3$$

The accurate answer is 2.899 which, rounded to two decimal places, is 2.90, so this is a good estimate.

b Estimate the result of 583.48 divided by 1.94.

$$\frac{583.48}{1.94}$$

Step 1: round the numerator (583.48) to the nearest whole number.

Numerator = 600

Step 2: round the denominator (1.94) to the nearest whole number.

Denominator = 2

Step 3: divide the numerator by the denominator.

$$\frac{600}{2} = 300$$

The accurate answer is 300.763 which, rounded to two decimal places, is 300.76, so this is a good estimate.

B Guided questions

Copy out the workings and complete the answers on a separate piece of paper.

1 **Estimate the total score from a study in which 185 participants completed a survey and the average score was 4.3. Show your working and check your answer.**

Step 1: round up 185 participants to 200 (the nearest hundred).

Step 2: round down average score to 4.

Step 3: multiply 200 × 4 = _____

Now work out the exact answer. Is it close to your estimate?

2 **Estimate the amount of time needed to process 130 participants when each participant procedure takes about 17 minutes. Express the answer in hours. Show your working and check your answer.**

Step 1: round down 130 participants to 100 (the nearest hundred).

Step 2: round up time needed for each participant from 17 to 20 minutes (the nearest ten).

Step 3: multiply $100 \times 20 = 2000$ minutes

Step 4: divide by the number of minutes in an hour = _____

Step 5: round up to nearest hour = _____

Now work out the exact answer. Is it close to your estimate?

3 **Estimate the number of students who will be awarded an A grade in a mathematics exam when there are 58 000 entries and 18.9% gain an A grade. Show your working and check your answer.**

Step 1: round up the number of students to the nearest easy number.

Step 2: round up the percentage who will be awarded an A grade to the nearest easy number.

Step 3: multiply the rounded quantities together.

Now work out the exact answer. Is it close to your estimate?

ⓒ Practice question

4 Henry was conducting research into generosity. He asked 18 participants to rate their own generosity on a scale of 1–20 (20 meaning extremely generous) and then asked the best friend of each participant to rate their friend's generosity using the same scale. The summary data from the investigation found:

Table 1.14

	Self rating of generosity	Friend's rating of generosity
Total	191	282

a Estimate the mean (average) for 'self rating' of generosity.
b Estimate the mean (average) for 'friend rating' of generosity.

Significant figures

> **REMEMBER**
>
> = means 'is equal to'
>
> > means 'is greater than'
>
> < means 'is less than'

Psychologists often show the data they have collected in tables and usually this data is shown to two or three significant figures. Because you could be asked to express the result of a calculation to two or three significant figures, you need to be able to use an appropriate number of significant figures and/or be able to calculate significant figures.

Significant figures in decimals

- 15.787 rounded to one decimal place (d.p.) = 15.8 because in 15.78**7 8 is > 5 so rounds up to .8**
- 3.533 rounded to one decimal place (d.p.) = 3.5 because in 3.5**33 3 is < 5 so rounds down to 3.5**

However, when numbers are tiny, for example, 0.00534 and 0.00328, rounding to one decimal place is not useful.

- 0.00534 rounded to one decimal place = 0.00
- 0.00328 rounded to one decimal place = 0.00

This is not very accurate as these two numbers, though small, are not the same!
So, to find an approximate answer with small numbers you **use significant figures**.

Counting significant figures

Significant figures start at the first non-zero number, so you ignore the zeroes at the front but not the zeroes between the significant figures!

Examples

- In 0.0093, the first significant figure is 9, the second significant figure is 3.
- In 0.0892, the first significant figure is 8, the second significant figure is 9, the third significant figure is 2.
- In 0.0706, the first significant figure is 7, the second significant figure is 0, the third significant figure is 6.

Worked examples

How many significant figures do the following numbers have?

a 0.**5007**   four significant figures

b **4.03**   three significant figures

c 0.00**3053**   four significant figures

d 0.0**745**   three significant figures

e 0.000**33**   two significant figures

Rounding significant figures

To round to a specific number of significant figures (s.f.) you use a similar method for rounding to a specific number of decimal places. You look at the number 'after' the one you are interested in to see whether it is greater or less than 5. If the number is equal to or greater than 5, then round up. If the number is less than 5, then round down.

 Worked examples

a **Round 0.0824591 to three significant figures (s.f.).**

   To round to three significant figures, look at the fourth significant figure. It is 5 so round up and the answer is 0.0825.

b **Round 0.5400205 to four significant figures (s.f.).**

   To round to four significant figures, look at the fifth significant figure. It is 2 so round down and the answer is 0.5400. Note, even though 0.5400 is the same as 0.54, include the zeroes to show that you have rounded to four significant figures.

B Guided questions

Copy out the workings and complete the answers on a separate piece of paper.

1 **Round 0.8356 to 3 s.f.**

   Step 1: identify the third s.f. as 0.83**5**6.

   Step 2: round UP the 5 because the following number (6) is greater than 5.

   The answer is _____

2 **Round 1.464 to 2 s.f.**

   Step 1: identify the second s.f. as 1.**4**64.

   Step 2: round UP the 4 because the following number (6) is greater than 5.

   The answer is _____

3 **Round 88.363 to 3 s.f.**

   Step 1: identify the third s.f. as 88.**3**63.

   Step 2: round your answer if necessary.

   The answer is _____

4 **Round 0.05687 to 3 s.f. Show your working.**

Step 1: identify the third s.f. _____

Step 2: round your answer if necessary.

The answer is _____

5 **Round 33.333 to 4 s.f. Show your working.**

Step 1: identify the fourth s.f. _____

Step 2: round your answer if necessary.

The answer is _____

C Practice questions

6 Write 0.700 as two significant figures.

7 Write 5.548 as four significant figures.

8 Write 6.683 as three significant figures.

9 Write 0.564 as two significant figures.

Levels of measurement

Note: this topic is assessed at AS level by OCR only.

Psychological researchers 'measure' behaviour and gather data and there are different levels of measurement. You need to be able to identify different levels of measurement so that you can state what level of measurement has been used in a study. You also need to be able to identify different levels of measurement because the level of measurement will determine how data is analysed. Make sure you can distinguish between levels of measurement.

Nominal level data

Nominal level data is a count of how frequently something occurs.

For example, count how many people in the class are left or right handed; count how many front doors in your street are red; count the frequency of blue-eyed or brown-eyed people in class. Nominal data is categorical data. So, if you are observing categories of 'things', usually a member of one category cannot also be a member of another category. For example, you are either left or right handed, a car is either blue or red, an animal is either a cat or a dog.

If research collects nominal level data, then you should be able to calculate ratios and percentages.

Guided question

Copy out the workings and complete the answers on a separate piece of paper.

1 **From an observational study, data found that 50 people chose an apple, 35 people chose a banana and 15 people chose a pear.**

   a **What is the ratio of apples to pears chosen? Simplify your answer.**

      Step 1: 50 people chose an apple and 15 people chose a pear so the ratio is 50:15.

      Step 2: 50 and 15 are both divisible by 5, so the simplified ratio is _____.

b   What is the ratio of apples to bananas chosen? Simplify your answer.

Step 1: _____

Step 2: _____

c   What percentage of people chose a pear? Show your working.

Step 1: 15 people chose a pear divided by the total number of observations of 100 (50 apples + 35 bananas + 15 pears) = 0.15

Step 2: multiply 0.15 by 100 = 15

Step 3: round up the answer to two decimal places so the answer is _____.

d   What percentage of people chose a banana? Show your working.

Step 1: _____

Step 2: _____

Step 3: _____

Ⓑ Practice question

2   From an observational study, data found that 18 people chose the red roses, 25 people chose the yellow roses and 22 people chose the pink roses.

   a   What is the ratio of red roses to pink roses chosen? Simplify your answer.
   b   What is the ratio of yellow roses to pink roses chosen? Simplify your answer.
   c   What percentage of people chose pink roses? Calculate the answer to two decimal places and show your working.
   d   What percentage of people chose red roses? Calculate the answer to two decimal places and show your working.

Ordinal level data

Ordinal level data is easy to remember because it means data (sets of scores) that can be put in order (ranked) lowest to highest, or highest to lowest, first score, second score, third score, etc.

For example, the scores in Data set X are put in order (ranked) lowest to highest, and the scores in Data set Y are ranked highest to lowest.

Data set X: 1, 3, 4, 6, 7, 9, 13, 21   Data set Y: 48, 33, 31, 25, 23, 21, 19

If research collects ordinal level data, then you should be able to calculate ratios and percentages.

Ⓐ Practice questions

1   Eight students sat an examination marked out of 50. Their scores were 22, 45, 34, 33, 48, 28, 37, 41.
   a   Put these scores in order, lowest to highest.
   b   Calculate the total of the scores.

c Calculate the percentage mark for the student who has the lowest mark. Show the answer to two decimal places.

d Calculate the percentage mark for the student who has the highest mark. Show the answer to two decimal places.

Interval level data

Interval level data is data that has been measured using a fixed scale. For example, time in minutes and seconds, weight in kilograms or height in centimetres.

For example, a cognitive psychologist wanted to find out whether there is a relationship between the amount of time people sleep at night and their reaction time, so the researcher collected interval level measurement because 'length of time slept' was measured in hours and minutes and 'reaction time' was measured in seconds.

If research collects interval level data, then you should be able to calculate percentages.

 Practice question

1 Ten students were competing to be chosen to represent their school in the 400 metre track event. The fastest student, Sam, ran 400 m in 69 seconds. The slowest student, Maddie, ran 400 m in 115 seconds.

Assuming they ran at the same speed:

a How long would Sam take to run 800 m in minutes and seconds?

b How long would Maddie take to run 800 m in minutes and seconds?

c As a percentage, how much longer did Maddie take to run 400 metres than Sam?

Examples of levels of measurement

Table 2.1

Source of data	Level of measurement	Explanation
Time taken to run 100 metres	Interval	Time is measured on a fixed scale.
An observation of how many cars stop at an amber light	Nominal	Observation of frequency of occurrence.
Examination scores from 20 students (maximum score = 100)	Ordinal	Scores can be put in order (ranked) high to low or low to high.
How many of a class of 20 students were awarded an A grade in an examination	Nominal	Observation of frequency of occurrence.
How many words beginning with 'S' participants can say in 30 seconds	Ordinal	Observations can be put in order (ranked) high to low or low to high.
Height of 20 students	Interval	Height is measured on a fixed scale.

A Guided questions

Copy out the workings and complete the answers on a separate piece of paper.

The examples below show how data is derived from six scenarios. The first three are explained for you. Complete the other three and check your answers.

1 **The data is the time taken for participants to complete a memory test.**

   This is **interval level data** because time is measured on a fixed scale: hours, minutes, seconds, etc.

2 **The observer counted the number of students who wore red on Tuesday.**

   This is **nominal level data** because it is data that can only be counted – so the measurement is frequency of occurrence.

3 **The survey counted the number of participants who said YES they did own a pet.**

   This is **nominal level data** because it is data that can only be counted – so the measurement is frequency of occurrence.

4 **The data is the score out of 20 for two groups who undertook a psychology test in either a warm room or a cold room.**

   Step 1: identify the level of measurement used.

   Step 2: explain why you have chosen this level.

5 **The data is the number of people who said YES (as opposed to NO) when asked to sponsor a charity event.**

   Step 1: identify the level of measurement used.

   Step 2: explain why you have chosen this level.

6 **The scores are the length of time (in minutes and seconds) each participant took to run 100 metres.**

   Step 1: identify the level of measurement used.

   Step 2: explain why you have chosen this level.

B Practice question

7 For each part of the question, identify the level of measurement and briefly explain why you chose this level of measurement.
   a The amount of time it takes for participants to react to a sound stimulus.
   b An observation in a supermarket looking at how many males and how many females buy a lottery ticket.
   c In a study of obedience, 26 of 40 participants administered a 450 V electric shock.
   d A cognitive psychologist showed one group of participants 20 pictures and another group 20 words and then both groups wrote down what they remembered. One point was awarded for each correct answer.
   e A group of 10 male and 10 female students are given 20 easy mathematics problems (addition and subtraction) to see how many they can solve correctly in 5 minutes.
   f Participants self-report the length of time (in hours and minutes) they slept the previous night.

Measures of central tendency: mean, median and mode

Measures of central tendency are used to summarise large amounts of data into typical or average values. You need to be able to explain the difference between the mean, median and mode, and be able to suggest which measure of central tendency may be the most appropriate for a given set of data and explain why this measure is appropriate.

A measure of central tendency is a central or 'typical' value of a set of scores. Measures of central tendency are often called averages. The most common measures of central tendency are:

- The arithmetic mean, which is the sum of all measurements divided by the number of scores in the data set.
- The median, which is the middle value that separates the higher half from the lower half of the data set (which must be ordered lowest to highest).
- The mode, which is the most frequent value (typical value) in a set of scores. This is the only central tendency measure that can be used with nominal level data.

A central tendency is usually calculated for a set of scores so that researchers can analyse how quantitative data scores cluster around some central value. There are three ways to calculate the central point of a set of scores: the arithmetic mean, the median and the mode.

The arithmetic mean

To calculate the arithmetic mean, all the scores are added up and the total is divided by the number of the scores. Here are two examples.

1 In this set of ten scores: 1, 3, 5, 3, 3, 4, 5, 5, 7, 8
   - The sum of the scores $(1 + 3 + 5 + 3 + 3 + 4 + 5 + 5 + 7 + 8)$ is 44, so the **mean** score is 44 divided by 10 which is 4.4.
2 In this set of 18 scores: 5, 6, 9, 7, 8, 9, 5, 6, 9, 7, 8, 9, 5, 6, 9, 7, 8, 9
   - The sum of the scores $(5 + 6 + 9 + 7 + 8 + 9 + 5 + 6 + 9 + 7 + 8 + 9 + 5 + 6 + 9 + 7 + 8 + 9)$ is 132.
   - The **mean** score is 132 divided by 18, which is 7.33333… This can be rounded to two decimal places to give 7.33.

Evaluation of using the mean as a measure of central tendency

The advantage of using the mean as a measure of central tendency is that the mean is a sensitive measure and it takes all the values from the raw scores into account.

The disadvantage of using the mean is that if there are unusual scores (very high or low outliers) the mean can be 'skewed' (pulled down or pulled up) to give a distorted impression. Also, because the mean is a statistic, if it is used inappropriately it may have decimal places that are not appropriate. For example, the mean may have a 'silly' decimal point – examples are families in the UK having 2.4 children, or owning 1.5 cats, or 2.2 cars.

The median

The median is the central score in a list of rank ordered scores. In an odd number set of scores, the median is the middle number. In an even number set of scores the median is the mid-point between the two middle scores. Here is an example.

In this set of ten scores: 2, 4, 4, 5, 5, 6, 7, 9, 15, 18
- The **median** is 5 + 6 divided by 2 which is 5.5.
- The **mean** of this set of scores is 75 divided by 10 which is 7.5.
- Note that in this example the mean is higher than the median.

Evaluation of using the median as a measure of central tendency

The advantage of using the median is that the median is not affected by extreme scores and it is useful when the data is ranked ordinal level data (1st, 2nd, 3rd, etc.).

The disadvantage of using the median is that the median only takes account of the position of the scores and does not take account of the values of the scores. It can be misleading if it is used in a small set of scores.

The mode

The mode (or modal score) is the score that occurs most frequently in a set of nominal level scores. Here is an example.

In this set of ten scores: 4, 4, 5, 5, 5, 6, 10, 12, 12, 14
- The **mode** is 5 because it occurs three times (the most frequently).
- The **median** of this set of scores is 5 + 6 divided by 2 = 5.5.
- The **mean** of this set of scores is 7.7 (77 divided by 10).

This example shows that each of the measures of central tendency may describe the middle of a set of scores differently.

Evaluation of using the mode as a measure of central tendency

The advantage of using the mode is that the mode is not affected by extreme scores and it is the only measure of central tendency that is usually appropriate to use with nominal level data.

The disadvantage of using the mode is that the mode tells us nothing about the other scores and there may be more than one mode in a set of data.

Worked examples

The examples below identify the most appropriate measure of central tendency for a given situation and explain the reasons for the choice.

a The data is the time taken to run 100 metres by 50 male and 50 female students.

As a measure of central tendency use the arithmetic mean. The data is interval level data, 50 is a large sample and the mean takes account of the value of each of the scores.

b An observation to find out whether pairs of males sit 'opposite' each other or 'side by side' in a library.

As a measure of central tendency use the mode. The data is nominal level data and the observed behaviour fits only into one of the two categories.

c A psychologist gave six students 30 seconds to write down as many parts of the brain (e.g. frontal lobe, amygdala) as they could remember. One point was awarded for each correctly named 'brain part'.

As a measure of central tendency use the median. The data is ordinal level data, the scores from each group can be ranked and the sample size (6) is small.

d A group of 10 male and 10 female students were given 20 simple mathematics problems (addition and subtraction) to see how many they could solve correctly in five minutes.

As a measure of central tendency use the median. The data is ordinal level data, the scores from each group can be ranked and the sample is small.

e How popular are types of flowers? An observation was carried out looking at the type of flowers bought in a supermarket on Saturday morning. The flowers available were daffodils, roses, lilies, tulips, daisies, mixed bunch.

As a measure of central tendency use the mode. The data is nominal level data and only the frequency of each category of flower sale is being measured.

 Guided question

Copy out the workings and complete the answers on a separate piece of paper.

1 Identify the most appropriate measure of central tendency for each situation. Give reasons for your choice.

a The height of 50 students.

As a measure of central tendency use the arithmetic mean because the data is interval level data, there is a large sample and the mean takes account of the value of each of the scores.

b An observation to find out whether two-year-old children play with others or play alone.

As a measure of central tendency use the mode because the data is nominal level data and the observed behaviour fits only into one of the two categories.

c In a fitness study, the average time it took for 200 participants to run 100 metres.

As a measure of central tendency use the arithmetic mean because the data is interval level data, there is a large sample and the mean takes account of the value of each of the scores.

Full worked solutions at www.hoddereducation.co.uk/essentialmathsanswers

d A teacher gave 10 students a spelling test of 20 words such as 'difficult', 'experiment', 'reliable', etc. One point was awarded for each correctly spelled word.

Suggest the measure to be used _____

Reason 1: level of data _____

Reason 2: size of sample _____

e A group of 10 male and 10 female students were given a five-minute test in which they were asked 20 simple questions about psychology. One mark was awarded for each correct answer.

Suggest the measure to be used _____

Reason 1: level of data _____

Reason 2: size of sample _____

f How popular are types of film? A survey asked people to identify the type of film they most prefer. The six film categories were 'romcom', 'horror', 'sci-fi', 'crime', 'period drama', 'comedy'.

Suggest the measure to be used _____

Reason 1: level of data _____

Reason 2: size of sample _____

Ⓒ Practice questions

2 The table shows two sets of scores from an experiment in which two groups of nine participants memorised and then recalled a list of 20 five-letter words, such as 'horse' and 'apple'. In the first condition the words were shown without pictures and in the second condition the words were shown with pictures.

Table 2.2

Without pictures	9	10	12	10	15	13	14	16	17
With pictures	10	11	13	14	16	16	17	15	16

a Put each set of scores into a table in order of lowest score to highest score.

b Calculate the median score for each condition:
  i Median score without pictures is _____
  ii Median score with pictures is _____

c Calculate the modal score for each condition:
  i Modal score without pictures is _____
  ii Modal score with pictures is _____

3 For each of the examples below, identify the appropriate measure of central tendency and explain why this measure of central tendency is appropriate.

a Twenty participants were observed as they chose a cake. They were offered five types of cake: chocolate, coffee, carrot, walnut, lemon drizzle.

b Average heart beat per minute taken from 10 students before and after they ran 100 metres.

c Fifty students were allocated to two groups and each group was given 20 short words to recall. One group recalled the words immediately and the other group chatted for five minutes before they recalled the words. The number of words recalled was recorded.

d Are people fitter in the morning? Twenty volunteers ran 100 m as fast as they could, first at 10 a.m. and then at 4 p.m. The time each participant took to run was recorded.

e Which is the most popular colour car? An observation was carried out in a large multi-story car park. Two observers watched each car entering the car park between 10 a.m. and 3 p.m. Their tally chart recorded seven colour categories: white, silver, red, black, blue, green, other.

Calculating the arithmetic mean

When you carry out practical research you will need to be able to calculate the arithmetic mean score (the average) for each of the conditions in your project. The arithmetic mean is the sum of a collection of numbers divided by the number of numbers in the collection.

> **REMEMBER**
>
> The algebraic equation for the arithmetic mean can be written as $\bar{x} = \frac{\Sigma x}{n}$
>
> where $\bar{x}$ represents the 'mean'; Σ represents 'the sum of'; x represents 'each of the scores' and n represents the number of scores.

The arithmetic mean is used to report the central tendency of a set of scores but, especially in small samples, it can be influenced by outliers. (Outliers are values that are much larger or smaller than most of the scores.) So, for skewed distributions, where a few scores are much greater, or smaller, than most of the scores, the arithmetic mean may not reflect the 'true' mid-point of the scores.

To calculate the arithmetic mean:

Step 1: add up all the scores to give a total score (T).

Step 2: divide the total score (T) by the number of scores.

Step 3: round to two decimal places.

A Worked examples

a **Calculate the arithmetic mean for the set of numbers 5, 8, 9, 8, 5, 6, 7, 6. Show your answer to two decimal places.**

Answer = 54 (the sum of the numbers) ÷ 8 (the number of the numbers) so the mean is 6.75.

b **Calculate the arithmetic mean for the set of numbers 21, 47, 36, 15, 41, 35. Show your answer to two decimal places.**

Answer = 195 (the sum of the numbers) ÷ 6 (the number of the numbers) so the mean is 32.50.

c **Psychologists used a questionnaire to find out the extent to which males and females enjoy shopping for clothes. The critical question was:**

'On a scale of 0–10, where 0 means "really dislike" and 10 means "really, really, enjoy", how much do you enjoy shopping?' The raw data is shown in the table.

Table 2.3

Participant	Male responses	Female responses
1	2	5
2	3	6
3	4	7
4	3	5
5	5	8
6	6	3
7	1	9
8	2	9
9	4	5
10	5	4
Total	35	61

Calculate the arithmetic mean of

i   the male responses

ii the female responses.

Show both answers to one decimal place.

i   Male mean = 35 (the total of all male responses) ÷ 10 (the number of male participants) = 3.5

ii Female mean = 61 (the total of all female responses) ÷ 10 (the number of female participants) = 6.1

B Guided questions

Copy out the workings and complete the answers on a separate piece of paper.

1 **Calculate the mean score: 20 + 14 + 18 + 35 + 25. Check your working and answer.**

There are five scores so calculate the mean by adding all of the scores to get the total and dividing by the number of scores.

$(20 + 14 + 18 + 35 + 25) \div 5$

The answer is _____.

2 **Calculate the mean score: 112 + 230 + 550. Check your working and answer.**

There are three scores so calculate the mean by adding all of the scores to get the total and dividing by the number of scores.

$(112 + 230 + 550) \div 3$

The answer is _____ and rounded to two decimal places is _____.

3 **Calculate the mean score: 140, 45, 90, 63, 22. Check your working and answer.**

Add all of the scores to get the total and divide by the number of scores.

The answer is _____.

4 **Calculate the mean score: 21.5, 55.3, 23.8, 18.5. Check your working and answer.**

Add all of the scores to get the total and divide by the number of scores.

The answer is _____ and rounded to two decimal places is _____.

C Practice question

5   A repeated measures experiment was conducted to find out whether stroking a pet affects how relaxed people feel. Eight volunteer participants agreed to self-rate how relaxed they felt, before and after they stroked a fluffy cat for 30 seconds. The critical question was 'How relaxed do you feel on a scale of 1–20, where 1 = very tense and 20 = very relaxed.'

Complete the table by calculating the total score for each condition and the arithmetic mean for each condition. Show the mean scores correct to two decimal places.

Table 2.4

Participant	Relaxation score before stroking cat	Relaxation score after stroking cat
1	14	15
2	12	16
3	15	18
4	9	17
5	11	13
6	8	14
7	16	14
8	10	9
Total		
Mean score		

Principles of sampling

When researchers conduct research, the **target population** is the group of people to whom they wish to generalise their findings. The **sample** of participants is the group of people who take part in the study and a **representative sample** is a sample of people who are representative of the target population.

Sample representativeness

Researchers wish to apply the findings of their research to explain something about the behaviour of the target population. Thus the sample of participants should be a true representation of diversity in the target population. In psychological research, students are often used as participants, but an all-student sample is only representative of a target population of students. Likewise, an all-male sample may only be representative of an all-male target population. If the sample is not representative, the research findings cannot be **generalised** to the target population.

Sampling techniques

When you design psychological research you have to decide how to gather a sample of participants from a designated population. There are several different ways to gather a sample.

Random sampling

This involves having the names of every member of the target population and giving everyone an equal chance of being selected. A random sample can be selected by a computer or, in a small population, by selecting names from a hat.

Strength: a true random sample avoids bias, as every member of the target population has an equal chance of being selected.

Weakness: it is almost impossible to obtain a truly random sample because, in a large population, all the names of the target population may not be known.

Opportunity sampling

This involves asking whoever is available and willing to participate. An opportunity sample is not likely to be representative of any target population because it will probably comprise friends of the researcher, or students, or people in a specific workplace. The people approached will be those who are local and available. A sample of participants approached 'in the street' is not a random sample of the population of a town. In a random sample, all the people living in a town would have an equal opportunity to participate. In an opportunity sample, only the people present at the time that the researcher was seeking participants would be able to participate.

Strength: the researchers can quickly and inexpensively acquire a sample and face-to-face ethical briefings and debriefings can be undertaken.

Weakness: opportunity samples are almost always biased samples, because who participates is dependent on who is asked and who happens to be available at the time.

Volunteer sampling

Volunteer samples are made up of people who 'select themselves' (volunteer) to participate. A volunteer sample may not be representative of the target population because there may be differences between the sort of people who volunteer and those who do not.

Strength: the participants should have given their informed consent, will be interested in the research and may be less likely to withdraw.

Weakness: a volunteer sample may be a biased sample of people who are not representative of the target population because volunteers may be different in some way from non-volunteers. For example, they may be more helpful (or more curious) than non-volunteers.

Systematic sampling

A systematic sample selects participants in a systematic way from the target population, for example, every tenth participant on a list of names. To take a systematic sample you list all the members of the population and then decide on a sample size. By dividing the number of people in the population by the number of people you want in your sample, you get a number (n) and then you take every nth member of the target population to get a systematic sample.

Strength: this method should provide a representative sample.

Weakness: systematic sampling is only possible if you can identify all members of the population to be studied.

Stratified sampling

In stratified sampling the researcher identifies the different types of people who make up the target population and works out the **proportions** needed for the sample to be representative.

A list is made of each variable of interest e.g. IQ, gender, age group, occupation, which might have an effect on the research.

For example, if you are interested in why people vote (or do not vote) for a political party, gender, age and income may be important, so you work out the relative percentage of each group in your population of interest. The sample must then contain all these groups in the same proportion as in the target population.

Strength: the sample should be highly representative of the target population and therefore you can generalise from the results obtained.

Weakness: gathering such a sample would be time consuming and difficult to do and this method is rarely used.

(A) Worked examples

a In a study of attitudes towards sport a psychologist put up a notice in a city sports centre advertising for participants. Which sampling technique was used?

This is volunteer sampling because people read the notice and then put themselves forward (self-selected) as volunteer participants.

b Are there gender differences in whether people consume alcohol in the street? To find out a psychologist carried out an observation in a town centre. Which sampling technique was used?

This is opportunity sampling. The people who are observed are not selected, they just happen to be in the town at the time of the observation.

(B) Practice questions

1 a A college principal wanted to find out why students did, or did not, choose to study mathematics. He put the name of every registered student into a computer program and the computer program selected 100 students to participate. Which sampling technique was used?

   b A fitness club was considering offering yoga lessons on Saturday morning so it decided to poll its membership. There were 5000 members of whom 45% were male and 55% were female. A sample of 45 males was selected from the male names and a sample of 55 females was selected from the female names. Which sampling technique was used?

2 If a psychologist selected a random sample of 30 students from a population of 250 students, what percentage of the population was selected?

3 A psychologist obtained a self-selected sample of 50 but halfway through the research four participants withdrew. What percentage of the original sample remained in the study?

4 A psychologist set out to obtain a snowball sample of people who were practising yoga. She asked a friend from her yoga class to participate and then her friend asked ONE friend and so on. Eventually 35 participants were selected and 40% of the

participants attended the same yoga class. How many of the participants attended the same yoga class?

5 A psychologist wanted to see if learning is influenced by class size. He gained permission from two schools to carry out research in their sixth forms.

In School 'A' sixth form there are 50 male and 75 female students and the average class size is 15. In School 'B' sixth form there are 100 male and 200 female students and the average class size is 30.

  a Calculate the ratio of male to female participants in School A. Simplify your answer.

  b Calculate the ratio of male to female participants in School B. Simplify your answer.

  c Identify the sampling technique that you would use to select a sample that is representative of the student population in School A.

  d Explain why you would use the sampling technique you identified in part **c**.

  e Describe how you would select a random sample of 25 participants from School B.

6 A psychologist wanted to see if age affects memory. She gained a volunteer sample of people. 25 were aged 18–20, 40 were aged 21–40 and 15 were aged 41–60.

  a Calculate the total number of participants.

  b Calculate the ratio of younger participants aged 18–20 to older participants aged 41–60. Simplify your answer.

  c Calculate the ratio of younger participants aged 18–20 to participants over the age of 21. Simplify your answer.

  d What percentage of the total participants were

    i aged 18–20

    ii aged 21–40

    iii aged 41–60?

Simple probability

Before you learn about inferential statistical tests, you need to understand simple probability. Probability is a measure of the likelihood that an event will happen. When dealing with probability, the outcomes of a process are the possible results. For example, when a dice is thrown the possible outcomes are 1, 2, 3, 4, 5 or 6 . In mathematical language, an event is a set of outcomes which correspond to the event happening. For instance, when throwing dice, throwing an even number is an event that corresponds to the set of outcomes 2, 4 or 6. The probability of an event, like throwing an even number, is the number of outcomes that constitute the event divided by the total number of possible outcomes.

Probability = the number of favourable outcomes ÷ the number of possible outcomes

Favourable outcomes

- If you 'bet' that when you throw a dice it will land on a 6, then there is only one favourable outcome, because only one of the six faces of the dice is a 6.
- If there are three horses in a race, horsey1, horsey2 and horsey3, and you 'bet' that horsey1 will win the race, then there is only one favourable outcome; that is that horsey1 will win.
- If there are five black horses in a race and three white horses and you 'bet' that a black horse will win the race, then there are five favourable outcomes, because there are five black horses in the race.

A Worked examples

a What is the probability of throwing a 5 if a dice is thrown once?

Throwing a 5 is an event with one favourable outcome (a throw of 5) and the total number of possible outcomes is six (1, 2, 3, 4, 5, or 6). Thus, the probability of throwing a 5 is $\frac{1}{6}$.

b What is the probability of throwing an even number?

Throwing an even number is an event having three favourable outcomes (2, 4 or 6) and the total number of possible outcomes is six (1, 2, 3, 4, 5 or 6). Thus, the probability of throwing an even number is $\frac{3}{6}$.

c If a coin is tossed twice, what is the probability that it will land as tails both times?

There is one favourable outcome (TT) and there are four possible outcomes (HH, HT, TH, TT) so the probability that the coin will land tails both times is one favourable outcome (TT) ÷ four possible outcomes (HH, HT, TH, TT) which is $\frac{1}{4}$.

Though these probabilities are calculated as fractions, they can be converted to percentages.

- In worked example **a** there is a 16.67% chance of throwing a 5.
- In worked example **b** there is a 50% chance of throwing an even number.
- In worked example **c** there is a 25% chance that the coin will land tails both times.

(For notes on Percentages see page 17.)

If all outcomes are favourable for a specific event, its probability is 1, for example, the probability of rolling a 6 or lower on one dice is $\frac{6}{6}$ or 1, or 100%.

If none of the possible outcomes are favourable for a specific event, such as throwing a 9 on one dice, the probability is $\frac{0}{6}$, or zero.

B Guided question

Copy out the workings and complete the answers on a separate piece of paper.

1 What is the probability of throwing a 3 if a dice is thrown once?

Step 1: identify the number of favourable outcomes.

Throwing a 3 is an event with one favourable outcome (a throw of 3).

Step 2: calculate the total number of possible outcomes.

The total number of possible outcomes is six (1, 2, 3, 4, 5, or 6).

Step 3: thus, the probability of throwing a 3 is _____.

C Practice questions

2 There are eight marbles in a bag. Two are red, one is green, one is yellow, one is black and three are blue. If you are asked to pull out a marble, what is the probability that you will pull out the black one?

3 There are 20 names written on tickets in a hat. Two of the tickets have your name on them. What is the probability that your name will be pulled out?

4 There are 5 red fish, 10 yellow fish and 20 blue fish. I select one fish at random. What is the probability that it will be a blue fish?

Probability as a 'level of confidence'

You can think of the probability of an event as the **level of confidence** that an event will happen. For example, the probability of throwing an odd number on one throw of the dice is an event having three favourable outcomes (possible outcomes are 1, 3 or 5) and the total number of possible outcomes is six (1, 2, 3, 4, 5 or 6). Thus, the probability of throwing an odd number is $\frac{3}{6}$, or 50%. So you can be 50% confident that an odd number will be thrown. Calculating the probability of an event is quite easy if the number of possible outcomes is known. However, this is not always the case.

Statistical significance

When psychologists test an alternative hypothesis (the alternative hypothesis is not the null hypothesis), for example, 'Group A will remember more words than Group B', even if the results show there is a difference, as scientists they must be **confident** that the difference is caused by the independent variable rather than being a 'chance' event. The independent variable (IV) is the variable that is manipulated in an experiment. The probability that the result of research (for example, a difference between two conditions) 'happened by chance' can be calculated and a minimum threshold of statistical significance can be set.

Probability (p) and levels of statistical significance

The **statistical significance** (the probability value or p-value) of a result indicates the degree to which the result is 'true' in terms of being representative of the population.

In other words, statistical significance refers to whether any differences observed between groups being studied (or relationships between variables) are 'real' or whether they are simply due to chance.

Researchers select the level of significance *before* conducting statistical analysis.

In psychology, usually either the 0.05 level (the 5% level of chance or 95% level of confidence) or the 0.01 level (1% level of chance or 99% level of confidence) is used.

If the probability of a chance result is less than or equal to the significance level (e.g $p \leq 0.05$) then **the null hypothesis is rejected** and the result is said to be statistically significant.

REMEMBER

The null hypothesis is what is tested.

Example

A researcher thought that students who are tested in the same room in which they learn will achieve higher marks on a test. Thirty students were given three minutes to memorise a list of 20 five-letter words beginning with P, such as 'plane', 'proud', 'poise', 'purse'. At the end of three minutes, 15 students were taken into another room and 15 remained where they were. All the students were given two minutes to write down as many words as they could remember.

You can write the null hypothesis in different ways.

- The number of five-letter words beginning with P recalled by students who are tested in the room where they learned them is not different to the number of words recalled by students who are tested in a different room.

- There is no significant difference in the number of five-letter words beginning with P recalled by students who are tested in the room where they learned them, compared to the number of words recalled by students who are tested in a different room.

- The location of the test, the same room or different room, in which the five-letter words are learned will have no effect on the number of words recalled by students.

If there is a difference between the two conditions, the difference may have been caused by the IV (in the example the test in the same room or different room) or just by chance. The statistical significance is the minimum level at which the null hypothesis can be rejected.

Usually, psychologists look for a probability equal to or less than 5% ($p \leq 0.05$) that the results are caused by chance, which means they can be at least 95% confident that the results are not caused by chance.

When you carry out a research project and find that the 'difference between two conditions are statistically significant', it means that you can reject the null hypothesis and be at least 95% confident your alternative hypothesis is true.

In terms of the null hypothesis, statistical significance is the minimum level at which the null hypothesis can be rejected. So if the statistical significance level is set at $p \leq 5\%$, and the probability that the outcome happened by chance is 2%, then the null hypothesis can be rejected.

A probability of $p \leq 0.01$ means that the probability is less than or equal to 1 in 100 (1%) that the results could have occurred *if the null hypothesis is true*. Therefore, the null hypothesis is rejected and the alternative (experimental) hypothesis is retained. (You can also say the alternative hypothesis is accepted.)

A probability of $p \leq 0.05$ means that the probability is less than or equal to 5 in 100 (5%) that the results could have occurred if the null hypothesis is true.

You must be able to explain the difference between the $p \leq 0.05$ and $p \leq 0.01$ levels of significance.

Table 2.5 Levels of significance

Level	Probability	Significance	When used
1% level	$p \leq 0.01$	highly significant	Where the researcher needs to be confident a null hypothesis is false. For example, 'Drug X has no effect on cancer.' Before prescribing Drug X a researcher should be at least 99% confident that this null hypothesis is false.
5% level	$p \leq 0.05$	significant	The conventional level for psychology research.
10% level	$p \leq 0.10$	marginal	When the researcher is not concerned about making a mistake. Perhaps because the research is a pilot study.

Type 1 and Type 2 errors

When psychologists decide whether to reject or retain the null hypothesis, they look at the results of a statistical test. However, there is always the possibility that they may make an error.

A **Type 1** error is deciding to reject the null hypothesis by concluding that the IV did have a significant effect on the dependent variable (DV) when actually the result was due to chance or some other factor.

A **Type 2** error is deciding to retain the null hypothesis by concluding that the IV had no significant effect on the DV when actually the result was caused by the IV.

> **REMEMBER**
>
> Type 1 or Type 2 errors can also be made when analysing the results of a correlation. A Type 1 error would be deciding that two variables are related when actually the relationship was due to chance or some other factor.

The level of significance (the probability) selected affects whether researchers are likely to make a Type 1 error or a Type 2 error. If researchers set the level of probability high at $p \leq 0.10$ they are more likely to make a Type 1 error. However, if researchers set the level of probability low, e.g. at $p \leq 0.001$, they are more likely to make a Type 2 error.

Guided questions

Copy out the workings and complete the answers on a separate piece of paper.

Use Table 2.6 to answer questions 1–3.

Table 2.6

Probability	Percentage/level of confidence
$p \leq 0.05$	95% Equal to or less than 5 chances in 100 that the result is due to chance.
$p \leq 0.01$	99% Equal to or less than 1 chance in 100 that the result is due to chance.
$p \leq 0.005$	Equal to or less than 5 chances in 1000 that the result is due to chance.
$p \leq 0.001$	Equal to or less than 1 chance in 1000 that the result is due to chance.
$p \leq 0.10$	90% Equal to or less than 10 chances in 100 that the result is due to chance.

1 **Two groups of students completed a spelling test. One group revised but the other group did not. If the difference between the scores of the groups was significant at $p \leq 0.10$, how many chances in 100 are there that it wasn't the revision that caused the difference?**

Step 1: what does $p \leq 0.10$ mean in terms of chances in 100 that the result is due to chance?

Step 2: with this level of confidence, what chance in 100 is there that revision caused a difference in the scores?

2 **If the findings of research are found to be significant at a probability of p ≤ 0.05, then what is the percentage level of confidence that the null hypothesis can be rejected?**

Step 1: what does $p \leq 0.05$ mean in terms of chances in 100 that the result is due to chance?

Step 2: what is the percentage level of confidence that the null hypothesis can be rejected?

3 **If the difference between Group A and Group B is found to be significant at a probability of p ≤ 0.01, then what is the percentage level of confidence that the null hypothesis can be rejected?**

Step 1: what does $p \leq 0.01$ mean in terms of chances in 100 that the result is due to chance?

Step 2: what is the percentage level of confidence that the null hypothesis can be rejected?

B Practice question

4 Researchers wanted to find out whether having an audience affects performance. Two groups of participants were given 100 coloured cards and asked to sort them into four colours (red, green, blue and black). Each participant in Group A sorted the cards on their own. Each participant in Group B sorted the cards while the rest of Group B watched. The time (in seconds) it took for each participant to sort the cards was recorded.

Table 2.7 shows the results.

Table 2.7

	Group A (no audience)	Group B (with audience)
Mean	25.5s	18.5s
Standard deviation	2.40s	3.60s

a The researchers found the results were significant at $p \leq 0.05$. What is meant by 'the results were significant at $p \leq 0.05$'?

b Further analysis found that the results were significant at $p \leq 0.01$. What is meant by 'the results were significant at $p \leq 0.01$'?

c Should the researchers reject the null hypothesis if the results are significant at $p \leq 0.05$? Explain your answer.

(For notes on Standard deviation see page 49.)

Measures of dispersion

Measures of dispersion show how spread out a set of data is. You have seen that measures of central tendency such as the median and mean represent the central value for a set of data. Within a set of data the raw scores differ from one another and also differ from the central (average) value. The extent to which the median and mean are good representatives of the values in the original set of scores depends upon the **variability** or **dispersion** in the data.

A set of scores has low dispersion when all the scores are similar to the mean and high dispersion when the set contains scores that are considerably higher and/or lower than the mean. You need to understand measures of dispersion because you could be asked to explain why measures of dispersion are useful and/or what a measure of dispersion suggests about a set of scores. Dispersion within a data set can be measured (or described) in several ways including the range and standard deviation. **(For notes on Standard deviation see page 49.)**

The range

The range is the easiest measure of dispersion to calculate and is the difference between the lowest and highest value in a set of scores. The range is useful for showing the spread within a set of scores and for comparing the spread between sets of scores.

Range is the difference between the highest and lowest scores. To calculate the range, subtract the lowest score from the highest score.

Some statistics books define range as the high score minus the low score, plus one. This is an inclusive measure of range, rather than a measure of the difference between two scores. For example: the inclusive range for data ranging from 6 to 10 would be 5. $(10 - 6 + 1)$

It doesn't matter which definition of range you use, as long as you are consistent, since you will only be comparing ranges measured the same way.

This book will define the range as the difference between the highest and lowest scores.

A Worked examples

a **If the highest value in a set of scores is 20 and the lowest value in the same set of scores is 15, calculate the range.**

The range is $20 - 15 = 5$

b **Calculate the range if the highest value in a set of scores is 55 and the lowest value in the same set of scores is 35.**

The range is $55 - 35 = 20$

c **Calculate the range if the highest value in a set of scores is 115.75 and the lowest value in the same set of scores is 98.25.**

The range is $115.75 - 98.25 = 17.50$

B Guided question

Copy out the workings and complete the answers on a separate piece of paper.

1 Table 2.8 shows the range calculated from the scores of two groups of participants who memorised and recalled 20 five-letter words in either silent or noisy conditions.

Table 2.8

Participant scores	1	2	3	4	5	6	7	8	9	10	Range calculation
Number of words recalled in silence	15	12	16	17	12	10	18	17	18	11	Highest score is 18 Lowest score is 10 **so range is 18 − 10 = 8**
Number of words recalled in noise	14	15	12	9	16	16	12	15	10	14	Highest score is 16 Lowest score is 9 **so range is 16 − 9 = 7**

C Practice questions

2 Calculate the range in each case.
   a The highest score is 120 and the lowest score is 40.
   b The highest score is 556 and the lowest score is 331.
   c The highest score is 95.5 and the lowest score is 35.25.
   d The lowest score is 1.25 and the highest score is 5.5.
   e The lowest score is 1 minute 30 seconds and the highest score is 3 minutes 10 seconds.

3 Table 2.9 shows how students performed in two mathematics tests. Calculate the range for both sets of scores

Table 2.9

Participant scores	1	2	3	4	5	6	7	8	9	10	Range
Score out of 50 in first mathematics test	45	39	25	33	40	29	42	37	46	26	
Score out of 50 in second mathematics test	39	41	28	35	42	34	38	35	41	27	

Normal and skewed distributions

In a set of scores, the scores are not always distributed evenly and you must understand the difference between a set of scores having normal distribution and a set of scores having a skewed distribution. You should also be able to look at the distribution of a given set of scores and indicate (estimate) the position of the mean (or median).

Normal distribution

In a set of scores having a normal distribution, most of the scores are near to the mean score and relatively few scores are much higher or much lower than the mean.

If a set of scores has normal distribution then the standard deviation (see page 49) can be used to determine the proportion of scores that lie within a specific range of the mean value.

In a normally distributed set of scores, it is **always** the case that:

- approximately 68% of scores are less than one standard deviation (1SD) away from the mean score
- approximately 95% of scores are less than two standard deviations (2SD) away from the mean score
- approximately 99% of scores are less than three standard deviations (3SD) away from the mean score.

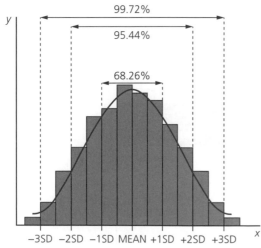

Figure 2.2 A frequency distribution with normal distribution.

Figure 2.2 shows the bell-shaped curve of normal distribution. The horizontal axis (x) shows the value of the score and the vertical axis (y) shows how many times that value occurred.

Full worked solutions at **www.hoddereducation.co.uk/essentialmathsanswers**

Normal distribution is a useful concept and you need to understand what normal distribution 'tells you' about a set of scores.

> **REMEMBER**
>
> In a normal distribution approximately:
>
> 68% of scores lie ± 1SD from the mean
>
> 95% of scores lie ± 2SD from the mean
>
> 99% of scores lie ± 3SD from mean

A Worked examples

a The mean of a normally distributed data set is 45 and its standard deviation is 4.5. What is the range of scores that lie within 1SD of the mean?

68% of the scores will lie between mean minus 1SD (45 − 4.5 = 40.5) and mean plus 1SD (45 + 4.5 = 49.5) thus 68% of scores will be between 40.5 and 49.5.

b In the same data set as in example a, what is the range of scores that lie within 3SD of the mean?

99% of the scores will lie between mean minus 3SD (45 − (3 × 4.5)) = (45 − 13.5 = 31.5) and mean plus 3SD (45 + (3 × 4.5)) = (45 + 13.5 = 58.5) thus 99% of scores will be between 31.5 and 58.5

c What percentage of the scores lie within 1SD above the mean?

68% of the scores will lie between mean minus 1SD and mean plus 1SD, thus 34% of scores will lie between the mean and 1SD above the mean.

d What percentage of the scores fall more than 2SD below the mean?

95% of the scores will lie between mean and 2SD above or below the mean, thus 2.5% will lie between the mean and 2SD below the mean.

B Guided questions

Copy out the workings and complete the answers on a separate piece of paper.

1 If there are 100 scores in a normally distributed set of scores, how many scores will lie within 1SD above the mean?

Step 1: what percentage of scores lie within 1SD of the mean in a normal distribution?

68% of the scores will be within ± 1SD from the mean.

Step 2: what percentage of scores lie within 1SD above the mean in a normal distribution?

Half of the 68% will be above the mean (the other half will be below the mean) so because there are 100 scores, 34% of the scores will lie within 1SD above the mean, thus 34 scores will fall within 1SD above the mean.

2 If there are 200 scores in a normally distributed set of scores, how many scores will be within 2SD below the mean?

Step 1: what percentage of scores lie within 2SD of the mean in a normal distribution?

95% of the scores will be within ± 2SD from mean.

Step 2: what percentage of scores lie within 2SD below the mean in a normal distribution?

C Practice questions

3 If there are 500 scores in a normally distributed data set, how many scores will lie within 1SD below the mean?

4 A researcher tells you that in his normally distributed data set 68% of the scores lie within 1SD above or below the mean. He tells you that 100 scores lie between the mean and 1SD above the mean. How many scores were there in the data set?

5 A researcher tells you that his data set comprises 200 scores and that 120 of the scores lie between the mean and 1SD above the mean. Is the data set normally distributed?

Skewed distributions

Data is not always normally distributed and the distribution of data around the mean may be skewed, meaning it has a long 'tail' on one side (of the mean) or the other. The skew may be positive or negative.

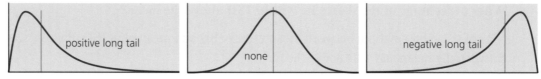

The red line indicates the position of the mean

Figure 2.3 Skewed frequency distributions

No skew

In the bell-shaped normal distribution, there is no skew and the mean is identical to the value at the highest peak of the distribution curve.

Positive skew

If data has a positive skew, then the long tail is on the right side of the peak and the distribution of scores can be said to be 'skewed to the right'. Thus **the mean is on the right and the mean value is higher than the value of the score at the peak of the distribution curve**.

Negative skew

If data has a negative skew, then the long tail is on the left side of the peak and the distribution of scores can be said to be 'skewed to the left'. Thus **the mean is on the left and the mean value is lower than the value of the score at the peak of the distribution curve**.

Causes of skew

Skewed distributions are caused by extreme values (outliers or anomalies).

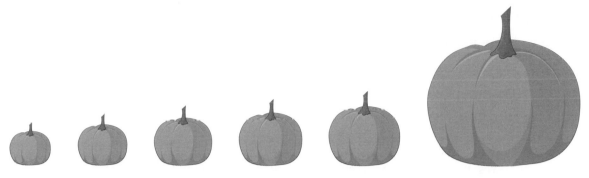

Figure 2.4

For example, in a competition to grow the biggest pumpkin, the total weight of the six pumpkins entered is 95 kg and the average weight of the pumpkins is 15.83 kg. The pumpkin weights are 6 kg, 7 kg, 8 kg, 9 kg, 10 kg and winner's weight is 55 kg. The outlier (the great big 55 kg pumpkin) has skewed the distribution. Five of the six data points (the pumpkin weights) are below the mean and because the mean is lower than the peak value of 55 kg, the data has a negative skew (a negative tail).

A Worked examples

a **Twenty students sat an examination marked out of 100. The marks awarded were: 1 at 20, 1 at 30, 8 at 40, 7 at 50, 1 at 90 and 2 at 100. Draw a chart to show the data and comment on the skew of the data.**

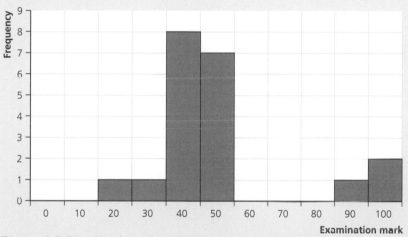

Figure 2.5 Distribution of marks in mock examination

There are two high value 'outliers'. The data has a positive skew as the mean is 'pulled to the right' by the two high scores.

The mean of the examination scores is 55.5. Have a look at the chart and see how most of the examination scores fall below the mean.

b **A researcher measured the height of ten children. They measured (in cm): 110, 120, 125, 126, 127, 128, 130, 121, 132 and 190. What can we say about this data?**

- The mean (average) height is 131.9 cm.
- Eight of the scores are below the mean.
- The outlier score is 190 cm.
- The data has a negative skew and the mean score will be to the left of the highest point in the skewed distribution curve.
- The range of heights is 190 − 110 = 80 cm.

B Guided question

Copy out the workings and complete the answers on a separate piece of paper.

1 **The examination scores of ten students were: 34, 44, 46, 55, 55, 55, 56, 56, 61 and 98.**

The mean score is 56 calculated as

$$\frac{\text{total score}}{\text{number of scores}} = \frac{560}{10}$$

a Is this data skewed?

Yes, because 8 of the 10 scores fall below (or are equal to) the mean score.

b What has caused the skew?

The high score (outlier) of 98 has pulled the mean score 'upwards'.

C Practice question

2 A teacher set 25 students a mock examination and the results are shown in Table 2.10.

Table 2.10

Examination score	Number of students
30	1
35	2
40	3
45	4
50	3
55	2
60	1
65	1
70	1
75	1
80	1
85	5

The frequency diagram shows the distribution of the scores.

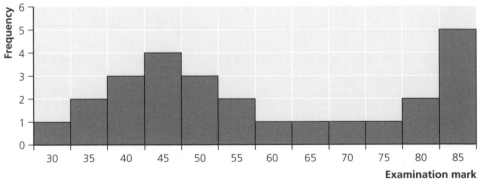

Figure 2.6 The distribution of marks in a mock examination

The mean exam score is $\frac{1435}{25} = 57.4$.

a Is this data skewed?

b What has caused the skew?

c Is the mean to the left or to the right of the highest point on the distribution curve?

d What is the modal examination score?

e What is the range of the examination scores?

Standard deviation

Standard deviation is a measure of dispersion that tells us the amount by which each score within a set of scores varies from the mean and thus how tightly the scores cluster around the mean. When the values in a data set are tightly clustered together (very little difference between the scores), the standard deviation is small (and of course the range will be small as well). When the scores are spread widely apart, the standard deviation is large. Standard deviation is the most useful measure of dispersion because, unlike the range, it takes into account the value of every score in the set of scores. In a set of scores having a normal distribution, most of the scores are near to the mean score and relatively few scores are extremely high or extremely low.

> **REMEMBER**
>
> In a normal distribution:
>
> 68% of scores lie within ± 1SD from the mean
>
> 95% of scores lie within ± 2SD from the mean
>
> 99% of scores lie within ± 3SD from the mean

Figure 2.2 on page 44 shows the concept of normal distribution. The horizontal axis (x) shows the value of the score and the vertical axis (y) shows how many times that value occurred.

How scores lie in a normal distribution

If the mean of a data set is 50 and its standard deviation is 5:

68% of the values in the data set will lie between mean minus 1SD ($50 - 5 = 45$) and mean plus 1SD ($50 + 5 = 55$).

99% of the values will lie between mean minus 3SD ($50 - (3 \times 5)) = (50 - 15 = 35$) and mean plus 3SD ($50 + (3 \times 5)) = (50 + 15 = 65$).

Ⓐ Guided question

Copy out the workings and complete the answers on a separate piece of paper.

1 **If there are 90 scores in a normally distributed data set, how many scores will lie within 1SD below the mean? Check your answer.**

Step 1: identify the percentage of scores that will be within ± 1SD from the mean. 68%

Step 2: half of the 68% will be below the mean (the other half will be above the mean) so 68% ÷ 2 = 34%.

Step 3: 34% of 90 scores is $\left(\frac{34}{100}\right) \times 90$ so 30.6 scores will lie within 1SD below the mean.

Calculating the standard deviation

There are two different calculations for the standard deviation and which formula you use depends upon whether the scores in your set of data represent an **entire population** or whether they are a **sample from a larger population**.

At AS and A-level you only need to learn how to calculate the standard deviation from a sample.

For example, if every student attending a college is asked how many times a week they eat in the college canteen then, because every student has been asked, the entire population has been studied. In such cases the population standard deviation should be used.

The formula to calculate the **standard deviation of a population σ (sigma)** is

$$\sigma = \sqrt{\frac{\Sigma(x - \mu)^2}{N}}$$

- x represents each value in the population of scores.
- μ is the mean value of the set of scores.
- Σ represents the total (sum of) the scores.
- N is the number of scores.

However, if a sample of 100 of the college students is asked how many times a week they eat in the college canteen, then only a sample of the population has been studied and the formula to calculate the **standard deviation of a sample s** is

$$s = \sqrt{\frac{\Sigma(x - \bar{x})^2}{n - 1}}$$

- x represents each value in the sample of scores.
- $\bar{x}$ represents the mean value of the sample of scores.
- Σ represents the total (sum of) the scores.
- $n - 1$ is the number of scores in the sample minus 1.

The **variance** is a measure of dispersion and is a mathematical expression of how scores (data points) are spread across a data set (sample or population). The variance is the mean of the squares of the distance each score in a set is from the mean of all the values in the set.

If, by now, you are starting to feel anxious don't worry because if you are asked to calculate the standard deviation of a set of scores in an examination you will be able to use a calculator.

A Worked example

Calculate the standard deviation given a small sample of five numbers: 5, 10, 15, 20 and 25.

The formula you need to use is

$$s = \sqrt{\frac{\Sigma(x - \bar{x})^2}{n - 1}}$$

Step 1: calculate the mean.

$$\text{Mean} = \frac{\text{total of the scores}}{\text{the number of scores}}$$

$$= \frac{(5 + 10 + 15 + 20 + 25)}{5}$$

$$= \frac{75}{5}$$

$$= 15$$

Step 2: calculate the variance.

Subtract the mean from each of the scores to calculate $(x - \bar{x})$.

$$5 - 15 = -10$$

$$10 - 15 = -5$$

$$15 - 15 = 0$$

$$20 - 15 = +5$$

$$25 - 15 = +10$$

Square all the answers you got from these subtractions to calculate $(x - \bar{x})^2$.

$$(-10)^2 = 100$$

$$(-5)^2 - 25$$

$$(0)^2 = 0$$

$$(5)^2 = 25$$

$$(10)^2 = 100$$

Add all of the squared numbers to calculate $\Sigma(x - \bar{x})^2$.

$$100 + 25 + 0 + 25 + 100 = 250$$

Divide the sum of squares by $(n - 1)$ to calculate $\sqrt{\frac{\Sigma(x - \bar{x})^2}{n - 1}}$.

n is the number of scores, so $n - 1$ is $5 - 1 = 4$, so the variance is $= \frac{250}{4} = 62.5$.

Step 3: to calculate the standard deviation find the square root of the variance.

$$\sqrt{62.5} = 7.905$$

Result! The standard deviation is 7.905. Rounded to two decimal places this is 7.91.

B Guided question

Copy out the workings and complete the answers on a separate piece of paper.

1 Calculate the standard deviation for the set of scores in Table 2.11.

Table 2.11

Participant	Number of words remembered	Score minus mean $(x - \bar{x})$	Score minus mean squared $(x - \bar{x})^2$
1	15	$15 - 14 = +1$	1
2	14	$14 - 14 = 0$	0
3	13	$13 - 14 = -1$	1
4	16	$16 - 14 = +2$	4
5	11	$11 - 14 = -3$	9
6	18	$18 - 14 = +4$	16
7	15	$15 - 14 = +1$	1
8	11	$11 - 14 = -3$	9
9	10	$10 - 14 = -4$	16
10	17	$17 - 14 = +3$	9
Total score	140		
Mean score	14		
Sum of squared differences			66

Step 1: calculate $\sum(x - \bar{x})^2$ for the data shown in the table (the sum of the squared differences between each score and mean). The calculations are shown in the table and the sum of the squared differences is 66.

Step 2: calculate the variance.

The sum of the squares is 66, so the variance is $66 \div (n - 1)$.

The number of scores is 10, so $n - 1 = 9$; thus the variance is $66 \div 9 = 7.3333\ldots$.

Step 3: calculate the standard deviation as the square root of the variance.

The standard deviation is $\sqrt{7.3333\ldots} = 2.7080$. Rounded to two decimal places this is 2.71.

C Practice question

2 Complete Table 2.12 to calculate $\sum(x - \bar{x})^2$ and then check your answer.

Table 2.12

Participant	Number of words remembered	Score minus mean $(x - \bar{x})$	Score minus mean squared $(x - \bar{x})^2$
1	21	$21 - 16 = +5$	25
2	22	$22 - 16 = +6$	36
3	18	$18 - 16 = +2$	4
4	12	$12 - 16 = -4$	16
5	11		
6	18		
7	15		
8	16		
9	12		
10	15		
Total score	160		
Mean score	16		
Sum of squared differences			

How is standard deviation useful?

Pierre, who is a French teacher, set two classes of students a test and then he compared the test scores. He found that the mean test score for Class A was much higher than the mean test score for Class B so he concluded that the students in class A were much better at learning languages.

Pierre told his friend 'Class B are a waste of time – they are useless at French.'

His friend (who is a psychology teacher) looked at the data and said 'Hang on – first you need to compare the standard deviation of both sets of scores.'

Pierre wasn't keen to do this but together they calculated the standard deviation of the test scores for Class A and Class B and found that Class A had a much larger standard deviation than Class B.

'What does that mean', asked Pierre?

His friend explained: 'A bigger standard deviation for Class A tells you that there is one (or more than one) student in Class A whose test score is extremely high and that suggests that the mean test score for Class A is skewed upwards.'

If the data had been the other way round, and the standard deviation for Class B had been much larger than for Class A, it would suggest that that there is one (or more than one) student in that class whose score is extremely low and that the mean test score for Class B is skewed downwards.

From this example, you can see that looking at the standard deviation may point you in the right direction when you try to interpret data and understand research findings.

Guided question

Copy out the workings and complete the answers on a separate piece of paper.

1 **Two groups of psychology students were tested. One group was tested in a noisy classroom and one in a quiet classroom. The test comprised 20 questions on research methods. The results are shown in Table 2.13.**

Table 2.13

	Group A Tested in noisy condition	Group B Tested in quiet condition
Mean	12.5	17.5
Standard deviation	2.3	4.4

What conclusions might be drawn from the data? Refer to the means and standard deviations in your answer.

From looking at the data we can see that when people are tested in quiet conditions they score higher than when they are tested in noisy conditions. This is supported *because the difference in mean test scores* shows that students tested in quiet conditions scored, on average, 5 higher than those tested in noisy conditions. However, the examination scores in the noisy conditions *are less varied (dispersed/spread out)* than in the quiet conditions. The lower *standard deviation* in the noisy conditions suggests that individual performances were more similar to each other and/or all quite close to the average score of 12.5.

This answer would be awarded high marks because the conclusions in respect of both means and standard deviations are clearly presented. Look at the text in italics and see how the answer demonstrates understanding of both the mean and the standard deviation and how quoting data demonstrates understanding.

B Practice questions

2 Does age affect memory?

Two groups of participants were recruited. Group A were aged 18–20 and Group B aged 50–60. Each participant was tested separately and asked to participate in a Kim's game in which they looked at 20 small items on a tray (e.g paper clip, pencil, marble) for one minute and were then given one minute to write down each item they could remember. The results are shown in Table 2.14.

Table 2.14

	Group A aged 18–20	Group B aged 50–60
Mean	15.4	13.6
Standard deviation	2.6	5.5

What conclusions might be drawn from the data? Refer to the means and standard deviation(s) in your answer.

3 Ten participants listened to a piece of classical music and then rated how much they liked it on a scale of 1–60, where 1 indicated 'I really disliked it' and 60 indicated 'I really, really, liked it'.

The mean score was 28 and the sum of the squared differences (the variance) was 2028. Calculate the standard deviation of the scores.

Selecting an appropriate statistical test

Note: this topic is assessed at AS level by OCR only.

In order to analyse the data from research and to calculate the probability of the results being significant, psychologists use statistical tests (inferential tests). There are two types of tests and the test that is appropriate for use depends on the research method and design. You must be able to select a suitable inferential test for a given practical investigation and explain why the chosen test is appropriate.

In the A-level examination (for OCR students at AS level) you may be asked to identify when to use a specific inferential statistical test and to explain why you selected the test you suggest. You need to be able to demonstrate knowledge and understanding of how and why inferential tests are used.

> **REMEMBER**
>
> **A statistically significant difference (or relationship)** is one which is unlikely to have occurred by chance. When a result is unlikely to have occurred by chance, then researchers will reject the null hypothesis and retain the alternative hypothesis.

Using inferential statistical tests allow psychologists to find out whether their findings are significant. Many inferential statistical tests require that data follows a **normal distribution**. You might want to revise the topic on normal distribution on page 44.

Remember that in a normally distributed set of scores:

- 68% of scores lie within ± 1SD from the mean
- 95% of scores lie within ± 2SD from the mean
- 99% of scores lie within ± 3SD from the mean

Thus in a normally distributed set of scores only 5% of the scores will fall more than +2 or −2 standard deviations above or below the mean.

However, sometimes the sample size may be so small that it is difficult to know whether the scores (data) are normally distributed and sometimes the scores (data) are a **skewed distribution**.

There are two kinds of statistical tests:

- **Parametric tests** can only be used when data is normally distributed.
- **Non-parametric tests** are used when data is not normally distributed.

Which inferential test to use?

Note: only AQA specifies that students should know the difference between parametric tests and non-parametric tests.

The sign test

The sign test is a non-parametric inferential test of difference that is suitable for use with related data (repeated measures design) and where numeric data is collected at nominal (or categorical) level. The sign test examines the direction of difference between pairs of scores. Here is an example.

A psychologist wanted to find out whether watching a party political broadcast on TV changed people's attitudes to that political party. A sample of 6 'undecided voters' was used and a base line measure of attitude was taken using a rating scale question where 0 = very unfavourable and 10 = very favourable. The participants then watched the leader of the political party give a five-minute TV broadcast, after which they were asked the same attitude question.

The results are shown in Table 2.15.

Table 2.15

Participant	Attitude	Direction of difference
1	More favourable	+
2	More favourable	+
3	No change	Omitted
4	Less favourable	−
5	More favourable	+
6	More favourable		

In the results four of the six participants reported more favourable attitudes. However, as an inferential test the sign test is not a very powerful test as it is only useful for repeated measures design and it only takes into consideration the direction of the differences rather than the value of any differences.

Mann–Whitney U test

The Mann–Whitney U test is **a non-parametric test** and is a test of the significance of the difference between two conditions. It is suitable for use when an independent design has been used AND the level of data collected is 'at least' ordinal.

Unrelated T test

The Unrelated T test is a **parametric test** and is a test of the significance of the difference between two conditions. It is suitable for use when an independent design has been used AND the level of data collected is 'at least' ordinal AND the data is normally distributed.

Wilcoxon matched pairs signed ranks test

This is a **non-parametric test** of the significance of the difference between two conditions when a repeated measures design has been used AND the level of data collected is 'at least' ordinal.

Related T test

This is a **parametric test** and is a test of the significance of the difference between two conditions when a repeated measures design has been used AND the level of data collected is 'at least' ordinal AND the data is normally distributed.

Spearman's Rho (Rank Order) correlation coefficient

This test is used when a correlation between two independent variables is being analysed. Spearman's is a **non-parametric test** and calculates the correlation coefficient between ranked scores when both sets of scores are 'at least' ordinal data (can be placed in rank order high to low or low to high).

Pearson's R correlation coefficient

This is used when a correlation between two independent variables is being analysed. Pearson's is a **parametric test** and calculates the correlation coefficient between actual scores which must be 'at least' ordinal.

Chi-square test

This is a test of significance of association which is used when nominal level data (frequency data) has been collected. (There is an example of a Chi-square on page 62.)

Guided questions

Copy out the workings and complete the answers on a separate piece of paper.

1 A researcher carried out a laboratory experiment having an independent design. There were 20 participants in each of the two conditions. The resulting ordinal data was found to be normally distributed. Which inferential test should the researcher use to analyse the data?

   ■ The researcher wanted to find out whether the difference between the two experimental groups is significant.

   ■ For an independent design with normally distributed ordinal level data the Unrelated T Test (a parametric test) is appropriate.

2 A researcher carried out a laboratory experiment having a repeated measures design. There were 20 participants who took part in both conditions. The resulting ordinal data was not normally distributed. Which inferential test should the researcher use to analyse the data?

   ■ The Wilcoxon matched pairs signed ranks test (non-parametric) is the test to use when data is not normally distributed.

   ■ It is also the appropriate test to use when a repeated measures design is used and the level of data is 'at least' ordinal.

3 A researcher carried out a study looking for a relationship between hours of sleep and levels of concentration. There were 20 participants who completed a survey on sleep and then a short concentration test. Sleep was measured in hours and concentration scores ranged between 0 and 10. The data was found not to be normally distributed. Which inferential test should the researcher use to analyse the correlation?

   ■ The Spearman's Rho (Rank Order) correlation coefficient is used when a correlation between two variables is being analysed.

   ■ Spearman's is a non-parametric test that calculates the correlation coefficient between ranked scores when both sets of scores are at least ordinal data (can be placed in rank order high to low or low to high).

4 A researcher carried out a study looking for an association between parenting styles and generosity in children. Parenting styles was defined as relaxed or strict. Generosity was defined as 'would share' a toy or 'would not share' a toy. Fifty parents completed a questionnaire and then their son/daughter was observed playing with another child. Which inferential test should the researcher use to analyse the association?

   ■ The Chi-square test is a test of significance of association which is used when nominal level data (frequency data) has been collected.

   ■ The resulting Chi-square table might look like this.

Table 2.16

	Would share a toy	Would not share a toy
Relaxed parenting style		
Strict parenting style		

B Practice questions

5 A researcher carried out a laboratory experiment having an independent design. There were 15 participants in each of the two conditions. The resulting ordinal data was found not to be normally distributed. Which inferential test should the researcher use to analyse the data?

6 A researcher carried out a laboratory experiment having a repeated measures design. There were 30 participants who took part in both conditions. The resulting interval level data was normally distributed. Which inferential test should the researcher use to analyse the data?

7 A researcher carried out a study looking for a relationship between age and verbal ability. One hundred participants aged 40 to 70 completed a short verbal ability task on which they could score between 0 and 50. The data was found to be not normally distributed. Which inferential test should the researcher use to analyse the correlation?

8 A researcher carried out a study looking for an association between self-efficacy and a mock examination outcome. One hundred students completed questionnaires in which they rated their self-efficacy as high, medium or low and their mock examination outcome as good (grade A or B), middling (grade C) or weak (grade D or E).

a Sketch the Chi-square that might result from this investigation.

b Why is Chi-square the appropriate statistical test to use?

Using statistical tests

When you have selected an appropriate statistical test, you need to be able to use it. Usually you will use a calculator or computer to calculate the results of a statistical test, but with research using small samples it is possible to calculate the result yourself. It is not difficult to calculate the Mann–Whitney U test, the Wilcoxon matched pairs signed ranks test, the Spearman's Rho (Rank Order) correlation coefficient, or the Chi-square.

> The formulae to calculate the Mann–Whitney U test and the Spearman's Rho test are given in Appendix 2.

The worked example shows how to calculate a Wilcoxon matched pairs signed ranks test.

A Worked example

A researcher investigates whether preference for fizzy drinks was influenced by brand labels.

The two-tailed hypothesis is that there is a difference in participants' preferred cola drink when participants taste brand-named or un-named cola drinks. Ten students took part in a test in which they were invited to taste five samples of numbered cola drinks and then select the number of the cola they liked the most.

Figure 2.7

Note: This is a two-tailed hypothesis because the researcher did not predict what the effect of seeing the brand labels would be.

Then the same participants were invited to taste five samples of branded cola drinks and asked to select the number of the one they liked the most.

The participants did not know that the samples of cola were the same in both conditions.

The results of the cola taste investigation are shown in Table 2.17.

Pepsi cola   Coca-Cola   Diet cola   Lidl cola   Asda cola

Figure 2.8

Table 2.17

	Taste 1 Numbered cola preference	Taste 2 Branded cola preference	Difference between scores (column A)	Ranked differences (column B)
1	3	1	1 − 3 = −2	4.5
2	1	5	5 − 1 = +4	9
3	4	1	1 − 4 = −3	7
4	2	5	5 − 2 = +3	7
5	1	2	2 − 1 = +1	2
6	4	3	3 − 4 = −1	2
7	4	5	5 − 4 = +1	2
8	4	1	1 − 4 = −3	7
9	2	2	2 − 2 = 0	ignore
10	3	1	1 − 3 = −2	4.5

Calculate the Wilcoxon matched pairs signed ranks test.

Step 1: calculate the difference between the two scores by taking one from the other. *This is shown in the table in column A.*

Step 2: rank the differences giving the smallest difference rank 1 (ignore the positive and negative signs). Do not rank any differences that are zero – miss them out! If there are differences that are the same, add together the rank positions they would have and divide by the number of 'sharers'. *This is shown in the table in column B.*

Step 3: add up the ranks for the positive differences.

Step 4: add up the ranks for the negative differences.

Step 5: calculate the sum of ranks for the positive differences.

$9 + 7 + 2 + 2 = 20$

Then calculate the sum of the ranks for the negative differences.

$4.5 + 7 + 2 + 7 + 4.5 = 25$

Step 6: T is the number that is the smallest when the ranks are totaled (T may be positive or negative).

$T = 20$

Step 7: N is the number of scores when those with zero difference are ignored.

$N = 9$

Step 8: RESULT: the smallest value of T is 20 and N = 9.

Using statistical tables to determine significance

Note: this topic is assessed at AS level by OCR only.

Critical values in interpretation of significance

Statistical tables tell you whether or not a calculated value is (or is not) significant. For all of the different inferential tests, mathematicians have calculated the critical values for significance. When the inferential test has calculated the result, you compare the result against the table of critical values. You need to be confident that, if given a value, you can use statistical tables to say whether the value is significant.

A Worked example

In the investigation looking at the influence of brand labels on fizzy drink preference, the researcher had written a two-tailed hypothesis that the brand labels would make a difference and had set a probability level of significance at $p \leq 0.05$ (5% chance).

From the calculation of Wilcoxon matched pairs signed ranks test the smallest value of T is 20 and N = 9. Is the result significant?

- If you look up the calculated T (T = 20) for N (N = 9) sets of scores in the Wilcoxon critical values table in Appendix 2, you find that for a two-tailed hypothesis, at $p \leq 0.05$, the critical value for N = 9 is 5.
- To be significant at $p \leq 0.05$ (5% chance), the calculated value of T (the smallest T) must be smaller (less) than the critical value.
- In this case, the calculated T value is 20, which is greater than the critical value, so the difference is not significant.
- Thus the researcher should retain the null hypothesis that brand labels do not affect fizzy drink preference.

B Guided questions

Copy out the workings and complete the answers on a separate piece of paper.

The critical values at different levels of significance for Spearman's Rho correlation coefficient are shown in Appendix 2.

Also, Table 2.20 on page 61 shows the critical values at different levels of significance for Spearman's Rho correlation coefficient. For example, given 8 pairs of scores, if a standard level of $p \leq 0.05$ is selected (probability less than or equal to 5% chance), the value of the calculated correlation coefficient must be equal to or greater than 0.63 to be significant.

REMEMBER

A calculated correlation coefficient may be a positive or negative number. (e.g. +0.65 or −0.65). If the result is negative this indicates a negative correlation. If the result is positive this indicates a positive correlation.

When you compare the calculated value to the critical value in the table you IGNORE the sign. So if the correlation coefficient from 8 pairs of scores was −0.83, at $p \leq 0.05$ this is still GREATER than the critical value of 0.63.

An investigation looked at the relationship between income and happiness. The null hypothesis is that income is not related to happiness.

1 **For each of the examples, use Table 2.20 to find the critical value for the correlation coefficient. The first two examples have been completed for you.**

Table 2.18

Number of pairs of scores	Level of significance	Critical value
10	$p \leq 0.05$	0.58
8	$p \leq 0.01$	0.76
5	$p \leq 0.05$	
10	$p \leq 0.01$	
7	$p \leq 0.05$	
6	$p < 0.01$	

2 **For each of the examples, use Table 2.20 to suggest whether the researcher should reject or retain the null hypothesis. The first two have been completed for you.**

Table 2.19

Number of pairs of scores	Level of significance	Calculated correlation coefficient (R)	Critical value	Reject or retain null hypothesis
10	$p \leq 0.05$	0.55	0.58	Retain null because 0.55 is less than 0.58
8	$p \leq 0.01$	0.85	0.76	Reject null because 0.85 is greater than 0.76
7	$p \leq 0.05$	−0.84	0.67	
9	$p \leq 0.05$	0.55	0.60	
5	$p \leq 0.01$	−0.77	0.87	

Table 2.20 Values of the correlation coefficient for different levels of significance

N	$p \leq 0.1$	$p \leq 0.05$	$p \leq 0.02$	$p \leq 0.01$
1	0.99	0.99	0.99	0.99
2	0.90	0.95	0.98	0.99
3	0.81	0.88	0.93	0.96
4	0.73	0.81	0.88	0.92
5	0.67	0.75	0.83	0.87
6	0.62	0.71	0.79	0.83
7	0.58	0.67	0.75	0.80
8	0.55	0.63	0.72	0.76
9	0.52	0.60	0.69	0.73
10	0.48	0.58	0.66	0.71

Chi-square and degrees of freedom

Degrees of freedom are important in a Chi-square test because, when you have calculated a Chi-square value, you need to know the degrees of freedom before you can look up probability of significance.

Degrees of freedom (df) can be thought of as the number of scores that are free to vary.

For example, if Cassandra tells you that she rolled four dice, that the total was 21, and that she rolled a 6 on the first dice, a 5 on the second and a 4 on the third, then you know that the fourth dice must have been a 6, otherwise the total would not add up to 21.

In this example, three dice 'falls' are free to vary while the fourth is not, so there are three degrees of freedom. Usually, the degrees of freedom are equal to the number of observations minus one, so for a sample size of 10 there are 10 observations and the degrees of freedom are $10 - 1 = 9$.

For a Chi-square test, the degrees of freedom can be calculated as the number of cells you need to fill in before you can complete the rest of the grid because if you know the total for each column and row, you don't have 'freedom' when filling in the values in the cells.

A Worked example

In an investigation looking at whether people who eat at least '5 a day' (fruit and vegetables) are less likely to get colds and flu, 80 participants self-reported their fruit and vegetable intake and whether they had cold and/or flu. Calculate y and z.

Table 2.21

	Always eat 5 a day fruit/veg	Never eat 5 a day fruit/veg	Total
Had flu last year	5	25	30
Did not have flu last year	35	y	z
Total	40	40	80

- The total ROWS and COLUMNS must add up to 80 (the total observations).
- When the values for the cells in the ROW 'had flu last year' are known, and the values for the cells in the COLUMN 'always eat 5 a day fruit/veg' are known, then z must equal 50 and so y must equal 15 as shown in Table 2.22.

Table 2.22

	Always eat 5 a day fruit/veg	Never eat 5 a day fruit/veg	Total
Had flu last year	5	25	30
Did not have flu last year	35	15	50
Total	40	40	80

The formula for Chi-square degrees of freedom (df)

The degrees of freedom for a Chi-square are equal to 'the number of rows minus one' × 'the number of columns minus one'.

This is written as the algebraic equation **df = (R – 1) × (C – 1)**.

In the example 2 × 2 Chi-square grid df = (2 – 1) × (2 – 1) or 1.

Critical values for Chi-square test

Table 2.23 shows the critical values at different levels of significance for the Chi-square test of association (for a two-tailed test). For example, if the degrees of freedom are 3 (df=3) and if $p \leq 0.05$ is selected as the level of significance, the value of the calculated Chi-square must be equal to or less than 7.82 for the null hypothesis to be rejected.

Table 2.23 The critical values at different levels of significance for the Chi-square test of association (for a two-tailed test)

Degrees of freedom	$p \leq 0.10$	$p \leq 0.05$	$p \leq 0.01$	$p \leq 0.001$
1	2.71	3.84	6.64	10.83
2	4.61	5.99	9.21	13.82
3	6.25	7.82	11.35	16.27
4	7.78	9.49	13.28	18.47
5	9.24	11.07	15.09	20.52
6	10.65	12.59	16.81	22.46

B Guided questions

Copy out the workings and complete the answers on a separate piece of paper.

1 **Write the df equations for the following Chi-square examples. The first three have been completed for you.**

  a 2 × 2 Chi-square   df = (2 – 1) × (2 – 1) = 1 × 1 = 1

  b 2 × 3 Chi-square   df = (2 – 1) × (3 – 1) = 1 × 2 = 2

  c 3 × 2 Chi-square   df = (3 – 1) × (2 – 1) = 2 × 1 = 2

  d 4 × 2 Chi-square

  e 3 × 4 Chi-square

  f 5 × 3 Chi Square

2 **For each of the examples, is the calculated Chi-square smaller than the critical value?**

  Table 2.24

Significance level	Degrees of freedom	Calculated Chi-square	Significant
$p \leq 0.05$	3	10.5 greater than 7.82	No
$p \leq 0.05$	4	8.8 smaller than 9.49	Yes
$p \leq 0.01$	2	12.25 greater than 9.21	
$p \leq 0.01$	3	9.5 smaller than 11.35	

(C) Practice question

3 Each of the examples shows a calculated Chi-square value, the degrees of freedom and the selected level of significance. For each example, use Table 2.23 to find (look up) the critical value and explain whether the result is significant or not.

Table 2.25

	Calculated Chi-square value	Degrees of freedom	Level of significance
a	15	1	$p \leq 0.05$
b	2.5	2	$p \leq 0.05$
c	10	3	$p \leq 0.01$
d	12	4	$p \leq 0.05$
e	5	1	$p \leq 0.01$

Differences between quantitative and qualitative data

Quantitative data is collected as numbers and qualitative data is collected as text or descriptions. The previous topics have focused on quantitative data but you also need to be able to explain how qualitative data can be analysed and/or converted to quantitative data.

Quantitative data

Strengths: it is objective, matter of fact, precise, high in reliability, can be used to make comparisons and statistical analysis.

Weaknesses: it may lack or lose detail, often results from unrealistic settings.

Qualitative data

Strengths: it is rich and detailed, more likely to be collected in real-life settings, provide information on attitudes, opinions and beliefs.

Weaknesses: it may be subjective, matters of opinion rather than fact, can be imprecise and difficult to analyse.

Analysing qualitative data

Qualitative data might result from video or audio recordings or written notes and also from open questions asked in interviews or questionnaires. When analysing qualitative data researchers must avoid subjective or biased misinterpretations. Misinterpretation can be avoided by:

- using accurate language to operationalise the variables to be measured, for example, if observing pro-social behaviour, a qualitative description of helpfulness might be 'picked up a dropped package' (though counting the frequency of this would be quantitative data)
- using a team of observers who have verified that they have achieved inter-observer reliability
- converting qualitative data into quantitative data.

Converting qualitative data to quantitative data

One way to do this is to use a coding technique and when a sample of qualitative data is collected – for example, from the interviewee, from a TV programme or from the notes or recordings of an observation – **coding units** are identified in order to categorise the data. A coding unit could be specific words or phrases that are looked for (the operationalised definitions). The coding units may then be counted to see how frequently they occur. The resulting frequency of occurrence is a form of quantitative data. Coding qualitative data in the form of text requires several processes:

- Decide (define) what your selected codes are.
- Read through the text and mark (e.g. underline or highlight) sections of the text that match your selected codes.
- Sort the results into some sort of order, group similar codes together and refine the coding scheme.

Thematic analysis can also be used to analyse qualitative data and in thematic analysis text is analysed to find 'themes' within data. Themes are patterns across data that are associated with a specific research question and the themes become the categories for analysis. Thematic analysis is performed through the process of coding in six phases to create established, meaningful patterns. These phases are:

1. familiarisation with data
2. generating initial codes
3. searching for themes among codes
4. reviewing themes
5. defining and naming themes
6. producing the final report.

Perhaps the easiest way to convert qualitative to quantitative data is to use a frequency count.

A study by Manstead and McCulloch (1981) analysed qualitative data (using content analysis) to examine how men and women were portrayed in a sample of British television advertisements, to see whether men and women were depicted differently. One hundred and seventy advertisements were analysed by coding the attributes of the adult central figure(s). Among the coding categories used were:

- sex
- mode of presentation
- relationship to product
- role
- location
- product type.

The analysis found that males and females were portrayed in different ways. Women were significantly more likely (than men) to be shown as product-users, to be shown in dependent roles, to be shown at home and to appear in advertisements for domestic products.

Example

A researcher is interested in the reasons why people do not engage in physical exercise. She issued questionnaires in which there was a closed question 'Do you engage in physical exercise every day?' (YES/NO) and an open question 'Why don't you engage in physical exercise?' The researcher defined the initial coding categories as:

- lack of time
- body image concern
- lack of money
- lack of motivation
- other.

When analysing the responses to the open question, the qualitative answers were placed in the most appropriate category (or categories) and the frequency of the categories then counted.

Table 2.26

Lack of time	Body image concern	Lack of money	Lack of motivation	Other
40	12	24	20	4

Another way to convert qualitative data to quantitative data is to design a scale having qualitative categories that are also assigned quantitative values. An example is the Bogardus social distance scale (Bogardus E.S. 1925) which is a psychological scale that measures people's willingness to participate in social contact of varying degrees of closeness with members of other social or cultural groups. The scale presents people with a list of qualitative descriptors and asks them to indicate the extent to which they would accept a member of a different 'group'. Each selection is scored and a score of 1 indicates no social distance.

The qualitative descriptors are:
- as a close relative by marriage (score 1)
- as my close personal friend (score 2)
- as a neighbor on the same street (score 3)
- as a co-worker in the same occupation (score 4)
- as a citizen of my country (score 5)
- as a visitor in my country (score 6)
- would exclude from entry into my country (score 7)

This scale is said to measure of how much liking and/or sympathy members of one group feel for members of another group, where low scores indicate liking or sympathy and high scores indicate disliking or no sympathy.

(A) Practice questions

1 You are carrying out a project observing how people protect their personal space on a busy commuter train. Suggest three qualitative descriptors that you might code in this investigation.
2 You are carrying out a study to investigate gender differences in the type of body language people use while talking with groups of friends. Suggest three qualitative descriptors that you might code in this investigation.

Difference between primary and secondary data

Primary data is data that is collected by a researcher dealing directly with participants. Primary data can be collected by different methods, including observation, surveys, interviews, experiments and case studies. Primary data is more reliable than secondary data because the researcher knows its sources.

Secondary data is data collected from external (secondary) sources, including television, radio, internet, magazines, newspapers, reviews, research articles and stories told by others. With secondary data, issues such as validity and reliability occur as the researcher can be less confident of the accuracy of the source.

Ⓐ Worked examples

a **A researcher observed people making food choices in a cafeteria. Is the data primary or secondary data?**

This is primary data because the observer collected the data directly from the participants as they made their food choices.

b **A researcher carried out a laboratory experiment looking at the effect of noise on concentration. Is the data primary or secondary data?**

This is primary data because the researcher collected the data directly from the participants.

c **A researcher gave a questionnaire to 100 people asking them about the effect of the weather on their mood. Is the data primary or secondary data?**

This is primary data because the researcher collected the data directly from the participants who answered the questionnaire.

d **A researcher carried out an analysis of attitudes to a new TV show by looking at social media blogs and twitter. Is the data primary or secondary data?**

This is secondary data because the researcher gathered the data from secondary sources.

Ⓑ Practice questions

1 A researcher observed how people use their mobile phones. Is the data primary or secondary data?

2 A researcher carried out a laboratory experiment looking at obedience to immoral orders. Is the data primary or secondary data?

3 A researcher gave a questionnaire to 100 people asking them about their diet and exercise. Is the data primary or secondary data?

4 A researcher carried out an analysis of attitudes to political parties by looking at social media blogs and twitter. Is the data primary or secondary data?

Order of magnitude

You probably won't often have to make order of magnitude calculations in psychology but it is useful to know what 'order of magnitude' refers to.

Orders of magnitude are used to make very **approximate comparisons** and reflect very **large differences**.

- If two numbers differ by one order of magnitude, one is about 10 times larger than the other.
- If two numbers differ by two orders of magnitude, they differ by a factor of about 100.

■ Two numbers of the same order of magnitude have roughly the same scale. The larger value is less than 10 times the smaller value. For example, 5000 and 3000 have the same order of magnitude.

Orders of magnitude are written in powers of 10.

For example, the order of magnitude of 1500 is 3, since 1500 may be written as 1.5×10^3.

Notice that this is written using standard form and the order of magnitude of the number is the power of 10 in standard form.

The **order of magnitude of a number** is the number of powers of 10 contained in the number. For example, the order of magnitude of the number 5 000 000 is 6.

An **order of magnitude is an estimate** of a variable whose precise value is unknown rounded to the nearest power of ten.

An **order-of-magnitude difference** between two values is a factor of 10. For example, the mass of the planet Saturn is nearly 100 times that of Earth, so Saturn is *two orders of magnitude* more massive than Earth. Order-of-magnitude differences are called decades when measured on a logarithmic scale.

Table 2.27

In words (long scale)	In words (short scale)	Prefix	Symbol	Decimal	Power of ten	Order of magnitude
thousandth	thousandth	milli-	m	0.001	10^{-3}	−3
hundredth	hundredth	centi-	c	0.01	10^{-2}	−2
tenth	tenth	deci-	d	0.1	10^{-1}	−1
one	one	–	–	1	10	0
ten	ten	deca-	da	10	10^{+1}	1
hundred	hundred	hecto-	h	100	10^{+2}	2
thousand	thousand	kilo-	k	1 000	10^{+3}	3
million	million	mega-	M	1 000 000	10^{+6}	6

Ⓐ Worked example

It is estimated that the number of neurons in the cortex of each of these mammals is:

hedgehog	2.4×10^7
cat	3×10^8
horse	1.2×10^9
human	2×10^{10}

Expressed as an order of magnitude comparison:

a **The number of neurons in a cat brain is ONE order of magnitude greater than in a hedgehog brain.**

Hedgehog = 2.4×10^7 and cat = 3×10^8

b **The number of neurons in a human brain is TWO orders of magnitude greater than in a cat.**

Cat = 3×10^8 and human = 2×10^{10}

c **The number of neurons in a human brain is ONE order of magnitude greater than in a horse.**

Horse = 2×10^9 and human = 2×10^{10}

B Guided questions

Copy out the workings and complete the answers on a separate piece of paper.

1 The lack of essential mathematics skills can be a problem for medical staff when they are administering drugs to patients and inaccurate calculations are still a significant source of drug error. Being given too much, or too little, of a prescribed drug is a risk to patients.

One problem for clinicians is that when they have to calculate drug doses, often the weights (and volumes) of the drugs are not given in the same units.

With weights, the unit of measurement changes every thousand. For example, you need 1000 micrograms (mcg) to make 1 milligram (mg) and 1000 milligrams to make one gram.

1 milligram $= 1 \times 10^3$ micrograms because a microgram is three orders of magnitude smaller than a milligram.

1×10^3 milligram $= 1$ gram because 1 gram is three orders of magnitude greater than 1 milligram.

A nurse was supposed to give her patient 500 mcg (micrograms) of a drug, but instead administered 5 mg (milligrams).

a Has the patient been given too much or too little of the drug?

Too much because 1000 micrograms make 1 milligram and the patient should only have been given 500 micrograms but instead was given 5 milligrams which is the equivalent of 5000 micrograms

b In terms of order of magnitude, how great (or small) is the mistake?

The dose is one order of magnitude greater than required, because the dose should have been 5×10^2 micrograms but the patient was given 5×10^3 micrograms.

C Practice question

2 A pharmacist was measuring out drugs. Each tiny box should have contained 1 gram of a drug, but instead each box contained 10 000 milligrams.

a Does each box contain too much or too little of the drug?

b In terms of order of magnitude, how great (or small) is the mistake?

3 Algebra

Algebraic symbols

You need to be able to understand and use algebraic symbols, be able to substitute symbols for given values, or values for given symbols, and be able to express the outcome of an inferential test in conventional form by using the appropriate symbols.

The symbols you must learn to use are shown below.

Table 3.1

Symbol	Meaning	Example
$=$	equals	$2 + 2 = 4$
$\neq$	not equal to	$5 + 1 \neq 3$
$>$	greater than	$5 > 3$
$<$	less than	$20 < 30$
$\geq$	greater than or equal to	≥ 0.05
$\leq$	less than or equal to	≤ 0.05
∞	infinity	describes something *without any limit*
$\sim$	an approximation	an estimate
$\gg$	much greater than	$1\,000\,000 \gg 1$
$\ll$	much less than	$1 \ll 1\,000\,000$

Ⓐ Worked examples

Look at these statements and make sure you understand why they are true or false.

Statement		True or False
a	$26 > 20$	True
b	$55 < 100$	True
c	$100 \neq 100$	False
d	$50 \geq 25$	True
e	$500 \leq 1000$	True
f	If $x = 3$ and $y = 4$, then $x + y = 7$	True
g	$15 + 9 \neq 24$	False
f	$33 - 22 > 10$	True
g	$599 + 399 \sim 1000$	True
h	$2000 \leq 3000$	True
i	$5 \ll 5000$	True

Full worked solutions at **www.hoddereducation.co.uk/essentialmathsanswers**

B Practice question

1 Decide whether each statement is true or false.

a $55 > 20$

b $5000 < 7000$

c $666 \neq 666$

d $500 \geq 250$

e $333 \leq 400$

f If $x = 8$ and $y = 9$, then $x + y = 20$

g $11 + 99 \neq 100$

h $33 + 22 > 10$

i $599 - 199 \sim 400$

j $55 \leq 65$

k $100 \gg 0.00001$

Substituting numerical values into algebraic equations

Note: this topic is assessed at AS level by OCR only.

To substitute means to put one thing in the place of another. For example, in football the 'substitutes' sit and wait on the 'subs bench' in case they are chosen to play in the place of another player. Another example of substitution is in sugar-free drinks which are often sweetened by sugar substitutes such as aspartame. In algebra substitution means replacing the letters, or symbols, with numbers, such as '8' instead of 'x'. You need to be able to substitute numerical values into algebraic equations using appropriate units.

Examples

- What is the value of x if $3 + 4 = x$?   Answer: $x = 7$
- What is the value of x if $4 + 5 = x$?   Answer: $x = 9$
- If $n = 12$, what is the value of $n - 8$?   Answer: $12 - 8 = 4$
- If $n - 1 = 20$, what is the value of $\frac{(n-1)}{2}$?   Answer: $n - 1 = 20, \frac{20}{2} = 10$
- Substitute $x = 5$, $y = 8$ and rewrite the expression $x^2 + y^2$ Answer: $5^2 + 8^2$
- What is $x + y + 25$ when $x = 5, y = 15$?   Answer: $5 + 15 + 25 = 45$

You need to be able to insert the appropriate values from a given set of data into a formula for a statistical test. You have already learned how to do this when you inserted the values from a set of data to calculate the standard deviation.

REMEMBER

The equation to calculate the standard deviation for a sample of scores is

$$s = \sqrt{\frac{\sum(x - \bar{x})^2}{n - 1}}$$

s is the symbol for standard deviation (in a sample of scores).

x represents each value in the sample of scores.

$\bar{x}$ represents the mean value of the sample of scores.

$\sum$ represents the total (sum of) the scores.

$n - 1$ is the number of scores in the sample minus 1.

Example

To rewrite the standard deviation formula for when the number of participants is 50:

$$s = \sqrt{\frac{\Sigma(x - \bar{x})^2}{n - 1}} \qquad \text{becomes} \qquad s = \sqrt{\frac{\Sigma(x - \bar{x})^2}{49}}$$

The examples below show how the symbols in the equation are substituted with numbers.

- The number of participants in a study is 20   so $n = 20$
- The number of participants in a study is 100   so $n - 1 = 99$
- There are two groups of participants

  Group A contains 20 participants   so $n_A = 20$

  Group B contains 18 participants   so $n_B = 18$

  The mean score for a study is 33   so $\bar{x} = 33$

  The total of all scores in a study is 45.5   so $\Sigma x = 45.5$

Ⓐ Practice question

1 Identify what the algebraic symbols mean. The first two have been completed for you.

Symbol	What the symbol means
a x	Each value in the sample of scores
b n	The number of participants
c $\bar{x}$	
d Σ	
e $\geq$	
f $\leq$	

Substitution in the Chi-square formula

The Chi-square statistical test is an important and useful test. The Chi-square test is used when the level of measurement is nominal and the research is looking for an association between two (or more) categories of variable. The Chi-square tests for an association between two categorical variables. It is appropriate when the observations are independent of each other (each category can only 'fit' into one of the 'squares') and the number in each 'cell' must be 5 or above in a 2 × 2 Chi-square table.

The purpose of the Chi-square test is to find the difference between an observed frequency and expected frequency and its value can be calculated using a formula.

The Chi-square symbol is x^2 and the formula is:

$$x^2 = \Sigma \frac{(O - E)^2}{E}$$

where

O = Observed frequency
E = Expected frequency
Σ = Summation
x^2 = Chi-square value

Example

If the observed frequency (O) is 20 and the expected frequency (E) is 15, the Chi-square formula is re-written as follows:

$$x^2 = \sum \frac{(O - E)^2}{E} \qquad \text{becomes} \qquad x^2 = \sum \frac{(20 - 15)^2}{15}$$

Solving algebraic equations

Note: this topic is assessed at AS level by OCR only.

Calculate a Chi-square

To bring together the various elements of your learning so far, the guided question shows you how to calculate a Chi-square statistic. To do this you need to substitute numbers for symbols and carry out some calculations.

A Guided question

Copy out the workings and complete the answers on a separate piece of paper.

1 **A psychologist wanted to find out whether people who eat at least 'five a day' (fruit and vegetables) are less likely to get 'flu. She surveyed 80 participants and asked them to self-report their fruit and vegetable intake in the previous year and whether they had 'flu.**

The results are shown in Table 3.1. The table shows the simplest case of Chi-square, a 2 × 2 contingency table.

Table 3.1 The association between eating fruit and vegetables and having 'flu

	Always eat five a day fruit/veg	Never eat five a day fruit/veg	Total
Had 'flu last year	5	25	30
Did not have 'flu last year	35	15	50
Total	40	40	80

If you are testing the observed frequencies of the data there are four possible categories:

- **always eat five fruit/veg and had 'flu**
- **never eat five fruit/veg and had 'flu**
- **always eat five fruit/veg and did not have 'flu**
- **never eat five fruit/veg and did not have 'flu**

Calculate the expected frequencies.

- You expect 'by chance' that 25% $\left(\frac{1}{4}\right)$ of the participants will fall in each of the four categories.
- 25% of the number of participants is 20 ($\frac{80}{100} \times 25 = 20$) so if having 'flu is just 'chance', and is not associated with eating five fruit/veg each day, you expect to find 20 people in each category.

In Table 3.2:

- Colum A is the category.
- Column B is the observed frequency for the category.
- Column C is the expected frequency for the category.
- Column D is the difference between the observed and expected frequency ($O - E$).
- Column E is the squared difference ($O - E)^2$.
- Column F is the square divided by the expected frequency $\left(\frac{(O-E)^2}{E} \right)$.

The cell shaded in YELLOW shows the sum of those quantities to give Chi-square (x^2) so the calculated Chi-square value is 25.

Table 3.2 The 'step by step' Chi-square calculation

Column A	Column B	Column C	Column D	Column E	Column F
Category	O (observed frequency)	E (expected frequency)	$O - E$	$(O-E)^2$	$(O-E)^2 \div E$
Always eat five fruit/veg and had 'flu	5	20	$(5-20)$ $=-15$	225	**11.25**
Never eat five fruit/veg and had 'flu	25	20	$(25-30)$ $=5$	25	**1.25**
Always eat five fruit/veg and did not have 'flu	35	20	$(35-20)$ $=15$	225	**11.25**
Never eat five fruit/veg and did not have 'flu	15	20	$(15-20)$ $=-5$	25	**1.25**
Total	**80**	**80**			**25**

You looked at what this calculated value of Chi-square means in the topic on page 60 when you learned how to use statistical tables to determine significance.

B Practice question

2 For the Chi-square table shown, calculate the values for all the blank cells and then state the value of Chi-square.

Table 3.3

Category	O (observed frequency)	E (expected frequency)	O – E	(O – E)²	(O – E)² ÷ E
Always eat five fruit/veg and had 'flu	60	50			
Never eat five fruit/veg and had 'flu	70	50			
Always eat 5 fruit/veg and did not have 'flu	40	50			
Never eat five fruit/veg and did not have 'flu	30	50			
Total	200	200			

There is an old saying that 'a picture is worth a thousand words'. Mathematical and numerical information is much easier to interpret if it is displayed in graphical format or as a visual diagram. You need to be able to transfer a set of numerical data into a graph or chart. You also need to be able to interpret what a graph or chart tells you about the results of a study.

Frequency tables, diagrams, bar charts and histograms

Psychologists use graphs and diagrams and charts to show data in visual displays because information provided in graphs and charts is easier to understand. You need to be able to **translate information between graphical, numeric and algebraic forms.** You must be confident that you can interpret data presented in a visual diagram or graph and/or select and sketch an appropriate type of graph or visual display. This topic explains the difference between frequency tables, line graphs, bar charts and histograms.

Line graphs

A line graph can be used to represent data in list form. Here is an example.

A class of students were asked to memorise a list of 18 words and were then given one minute to write down as many words as they could remember. The teacher recorded how many times each word was remembered and whether the position of the word in the list would affect whether it was remembered. The results are shown in Table 4.1 and in the line graph in Figure 4.1.

The table shows the raw data. In the line graph:

- The x-axis (horizontal) is the position of the word in the list.
- The y-axis (vertical) is how many times that word was remembered.

Table 4.1

Position of word	Number of times each word was remembered
1	16
2	15
3	16
4	14
5	13
6	12
7	13
8	14
9	10
10	9
11	9
12	6
13	5
14	7
15	7
16	10
17	12
18	14

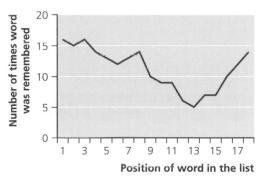

Figure 4.1 Results of a memory test when a list of 18 words was memorised

By looking at the line graph the teacher can see that the 1st, 2nd, 3rd and 4th words were recalled more often than the 12th, 13th and 14th words, and that the last few words on the list were remembered more than the words in the middle of the list. If you are studying memory, you will know this is called the primacy-recency effect (the serial position effect).

Pie charts

A pie chart is divided into sectors to illustrate numerical proportion. In a pie chart, the angle of each sector is proportional to the quantity it represents. Pie charts are widely used in the business world but it is difficult to compare different sections of a given pie chart, or to compare data across different pie charts. Pie charts can be replaced in most cases by other diagrams such as the bar chart.

Here is a pie chart representing the school examination results shown in Table 4.2.

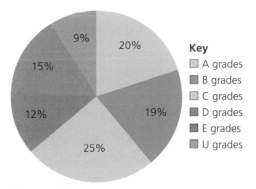

Key
- ☐ A grades
- ■ B grades
- ☐ C grades
- ■ D grades
- ■ E grades
- ☐ U grades

Figure 4.1

Table 4.2

A grades	20%
B grades	19%
C grades	25%
D grades	12%
E grades	15%
U grades	9%

To draw a pie chart, you need to represent each part of the data as a proportion of 360, because there are 360 degrees in a circle.

For example, if 25% of the 100 examination grades are C grades you will represent this on the circle as a segment with an angle of $\left(\frac{25}{100}\right) \times 360 = 90$ degrees.

Bar charts

Bar charts are used when scores are in categories (nominal level data), when there is no fixed order for the items on the y-axis. They can be used to show a comparison of means. The bars in bar charts should be the same width **but do not touch**. The space between the bars illustrates that the variable on the x-axis is **categorical or discrete data**.

For example, a psychologist investigating the types of pet people own carried out a survey of 200 families. Here are the results.

Table 4.3

Type of pet	Cat(s)	Dog(s)	Fish	Hamster	Rabbit	Other pet	No pet
Percentage of families	35	26	6	12	11	5	5

In the bar chart, the percentage of families is shown on the vertical (y) axis and the type of pet(s) shown on the horizontal (x) axis. By looking at the bar chart, it is easy to see that in this sample, the largest percentage of families owned a cat.

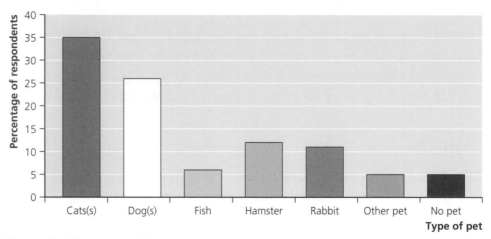

Figure 4.3 Pet ownership

Frequency tables

A frequency table is a table showing the values of one or more variables. The data collected during an observation is often collected on a frequency table. The 'tally chart' example is a frequency table from an observation of car colours.

Table 4.4 Frequency of observed car colours

Colour of car	Tally	Frequency
red	╫╫ ╫╫ I	11
green	╫╫ III	8
silver	╫╫ ╫╫ ╫╫ I	16
blue	╫╫ I	6
black	III	3
white	╫╫ ╫╫ II	12
other	IIII	4
total		60

Histograms

If the IQ scores from the population of the UK was presented as a distribution curve it would look like this.

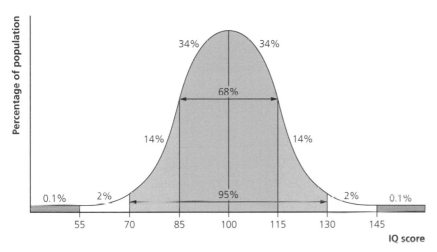

Figure 4.4 IQ score distribution

Histograms (frequency diagrams) show frequencies using columns. Histograms should be used to display frequency distributions of **continuous data** and there should be **no gaps between the bars**.

This histogram shows the distribution of height in a sample of 55 secondary school students.

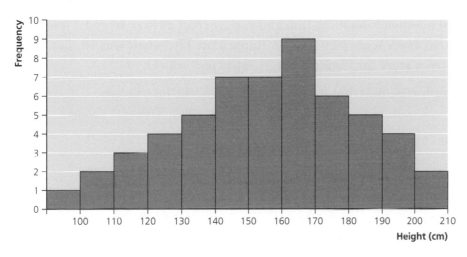

Figure 4.5 Height of a sample of secondary school students

A Worked example

Table 4.5 shows the average temperature recorded in Gtown each day of one week. Construct a bar chart to show the data in the temperature table.

Table 4.5

Week beginning 10 June	Monday	Tuesday	Wednesday	Thursday	Friday	Saturday	Sunday
Temperature (°C)	15	14	16	18	22	20	14

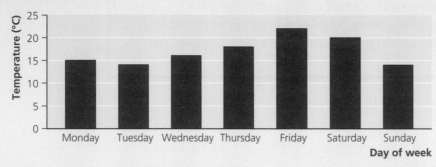

Figure 4.6 Average temperatures in Gtown week beginning 10 June

The example is good because:

- it is a bar chart
- axes are labelled
- there is an appropriate and informative title (in the caption)
- the y-axis has an appropriate scale
- the x-axis is labelled clearly
- the bars are accurate.

B Guided question

Copy out the workings and complete the answers on a separate piece of paper.

1 Participants in an experiment were shown a comedy film and then divided into two groups.

Group A were interviewed immediately. Group B underwent a delay condition and participants waited an hour before they were interviewed. Each participant was asked 20 questions about the film and each correct answer scored 1 point. The median accuracy score for Group A 'immediate interview' was 16. The median accuracy score for Group B 'delay interview' was 11.

Draw an appropriate diagram to show the results of the interviews.

You should:
- draw a bar chart
- label both axes
- give your diagram an informative title
- give the y-axis an appropriate scale
- separate the bars
- plot bars reasonably accurately.

Your graph should look very similar to the one on the following page:

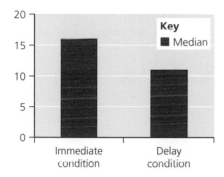

Figure 4.7 Scores for 'immediate' or 'delayed' interview

ⓒ Practice question

2 Between 2 p.m. and 3 p.m. on a Saturday afternoon researchers observed cars driving along a high street. They recorded the colour of each car they saw. Table 4.6 is the frequency table.

Table 4.6

Colour of car	Frequency
red	10
green	12
silver	25
blue	11
black	15
white	15
other	12
total	100

a Draw an appropriate graph to show the results of the observation.
b Explain why you chose this type of graph.
c What level of data is being collected?
d Which measure of central tendency should be used for this research?

Scatter diagrams
Plot two variables from an investigation

Another type of visual diagram is a scatter diagram. You need to **be able to plot two variables from an investigation** onto a scatter diagram and identify the pattern as showing a positive correlation, negative correlation or no correlation.

Scatter diagrams are used to depict the result of correlational analysis. You can usually see at a glance whether there appears to be a positive, negative or no correlation.

To draw a scatter diagram, a dot or cross is placed where the lines drawn from the x-axis and y-axis cross.

For example, in Figure 4.8, A is a scatter diagram showing no correlation, B is a scatter diagram showing a positive correlation and C is a scatter diagram showing a negative correlation.

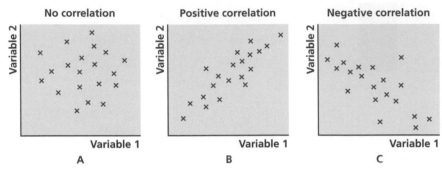

Figure 4.8

Ⓐ Worked example

Twenty students were asked to take part in a study looking at the relationship between quality of sleep and anxiety about examinations. Each day of a week while students were sitting examinations, the participants completed a questionnairc on which they reported how anxious they had been that day (on a scale of 1 to 10 where 1 = not at all anxious and 10 = very anxious) and how well they had slept the previous night (on a scale of 1 to 10 where 1 = not at all well and 10 = slept very well).

At the end of the week the researchers calculated the average anxiety score and the average sleep score for each of the students. The results are shown in Table 4.7.

Table 4.7

Anxiety score	Sleep score
6	4
7	4
9	5
8	5
5	7
4	7
6	7
6	5
7	5
9	4
5	6
5	9
3	8
2	8
7	2
7	3
8	3
9	3
6	3
5	4

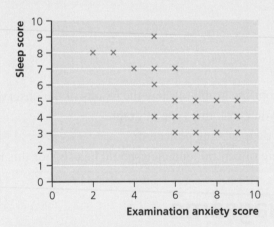

Figure 4.9 Sleep score correlated to examination anxiety

a **Does the scatter diagram show a positive, negative or no correlation?**

The scatter diagram shows a slight negative correlation between examination anxiety and sleep scores.

b **What level of measurement are the sleep scores and the anxiety scores?**

The level of measurement is ordinal as both sleep scores and anxiety scores can be ranked lowest to highest.

B Practice question

1 Ten students were asked to report how many hours of revision they did before a mathematics examination. Their marks (maximum 50) were correlated with the amount of time they revised. The results are shown in Table 4.8.

Table 4.8

Revision hours	Marks in mathematics examination
3	30
4	35
5	40
8	42
5	50
6	33
7	45
3	25
5	38
2	39

a What level of measurement are the mathematics examination marks?

b Draw a scatter diagram to show the relationship between the number of hours of revision and the examination marks.

c Comment on whether your scatter diagram shows a positive, negative or no correlation.

Exam-style questions

Regardless of which awarding body you are studying (AQA, OCR, Edexcel, or WJEC/ Eduqas) in the examination there will usually be a description of a hypothetical research study, the raw data that results from the study and/or some descriptive statistics, and you will be asked questions about the data, results and conclusions that can be drawn from the study.

The following questions are written to look like the type of questions you will find in examinations and you should revise thoroughly before attempting these questions. You will find suggested answers and guidance regarding how marks are likely to be awarded online at www.hoddereducation.co.uk/essentialmathsanswers.

> **REMEMBER**
>
> Questions about statistical tests are specified as being assessed at A-level only except by OCR.

1 Psychologists used a questionnaire to investigate the attitudes of local men and women to a planned housing development. They used a rating scale question from 0 to 10 and categorised scores of 0 to 5 as having a negative attitude and 6 to 10 as having a positive attitude.

Table E.1

Participant	Attitude score Male	Attitude score Female
1	5	6
2	6	6
3	4	5
4	4	5
5	5	6
6	5	4
7	7	7
8	7	8
9	4	9
10	3	7
Mean rating	5.0	6.3
Mode rating		

a   Complete the table to show the mode of the male and female attitude scores.   **(2)**
b   Give one reason why, when analysing this data, the mode is not the most useful measure of central tendency.   **(1)**
c   Identify one measure of dispersion that could be used to analyse this data.   **(1)**
d   Identify an appropriate statistical test that could be used to analyse this data.   **(1)**

e Explain why the test you selected in part **d** is appropriate to analyse this data. **(3)**

f Sketch an appropriate graph to illustrate this data. **(3)**

2 Mario was investigating memory. First he tested 10 students by putting 20 familiar objects, such as a pen, a text book, a pencil, on a table and asking them to study the objects for one minute and then giving them one minute to write down as many of the objects that they could remember. Then he tested the same 10 students again. This time there were 20 unfamiliar objects on the table, e.g. a cheese grater, a tea strainer, and again students studied the objects for one minute and were given one minute to write down all the objects they could recall. The results are shown in Table E.2.

Table E.2

	Memory for familiar objects	Memory for unfamiliar objects
Mean number of objects recalled	15	9

a Identify an appropriate statistical test that could be used to analyse this data. **(1)**

b Explain why the test you selected in part **a** is appropriate to analyse this data. **(3)**

c Sketch an appropriate graph to illustrate this data. **(3)**

3 A behaviourist researcher was studying learning in chickens. He put a chicken in a large cage and trained it to ring a bell. Each time the chicken pecked the bell, the bell rang and the chicken was rewarded with some seeds. At first the chicken pecked at other things and ignored the bell, but gradually the time between the pecks at the bell decreased. The data is shown in Table E.3.

Table E.3 Time taken for the chicken to ring the bell

Trial/Attempt	Time taken for the chicken to ring the bell (seconds)
1	120
2	65
3	45
4	33
5	28
6	22
7	15
8	9
9	5

a Calculate the mean time it took for the chicken to ring the bell. Show your calculations. **(2)**

b Which of the following is true?
The ratio between the 1st time to peck the bell and the 9th time to peck the bell is:
  A 12 times faster than it was at the start
  B 20 times faster than it was at the start
  C 24 times faster than it was at the start
  D 9 times faster than it was at the start **(1)**

c Calculate the range of the times the chicken took to ring the bell. Show your calculations. **(1)**

d Identify the level of measurement used to measure the chicken's learning. **(1)**

4 A psychologist was investigating whether the presence of others influences whether people laugh at jokes. In experimental research, in Condition A (the group condition) 10 participants sat and watched a film of a stand-up comedian performing 15 jokes. Four hidden observers noted each time a participant laughed aloud or smiled. In Condition B (the alone condition) another 10 participants watched the same film one at a time and a hidden observer noted each time the participant laughed aloud or smiled. The results of the study are shown in Table E.4.

Table E.4 The effect of social influence on whether people laugh at jokes

	Condition A (Watched as group)		Condition B (Watched as individual)	
	Laughed	Smiled	Laughed	Smiled
Mean	9.5	16.5	5.5	11.0
Standard deviation	1.25	2.5	3.6	4.85

a What conclusions might be drawn from the data? Refer to the mean and the standard deviation in your answer. **(6)**

b Identify the level of data that was collected in the study. **(1)**

c Name the statistical test you could use to analyse the difference between the scores and explain why the test you have selected is appropriate. **(4)**

d Briefly explain why standard deviation is a useful measure of dispersion. **(3)**

e The psychologist found the difference between the number of times people laughed in the group condition and in the alone condition was significant at $p \leq 0.05$. What is meant by 'the results were significant at $p \leq 0.05$'? **(2)**

5 An experiment was carried out to see whether people are more likely to remember information associated with their gender. Ten female and ten male students were given one minute to memorise a list of 20 words. Ten of the words were judged to be female related (e.g. lipstick, bikini, perfume) and 10 of the words were judged to be male related (e.g. football, shorts, shave). The participants were then given one minute to write down all the words they could recall.

Table E.5 shows the data for the female words recalled.

Table E.5 Number of female related words recalled

	Condition A (Male)	Condition B (Female)
1	4	6
2	5	6
3	4	7
4	6	8
5	6	6
6	7	6
7	5	5
8	8	8
9	2	5
10	1	6

a Identify the measure of central tendency you would use to find the average score and explain why you would choose this measure. **(2)**

b Name a statistical test you could use to analyse the difference between the scores and explain why the test you have selected is appropriate. **(3)**

c The psychologist found the difference between the number of words remembered in each condition was not significant at $p \leq 0.05$. What is meant by 'the results were not significant at $p \leq 0.05$'? **(2)**

6 Is there a relationship between teaching and mathematics confidence? Ten students who said they lacked confidence in their mathematical ability were invited to attend extra tuition sessions for two months. The number of tuition sessions each student attended was recorded and, at the end of two months, the students were asked to self-report their mathematics confidence on a scale of 1 to 10 where 1 was not at all confident and 10 was extremely confident. The results are shown in Table E.6.

Table E.6 Extra tuition attendance and mathematics confidence

Participant	Number of times attended extra tuition	Mathematics confidence
1	2	4
2	4	4
3	3	5
4	5	5
5	6	6
6	6	6
7	7	7
8	7	9
9	8	9
10	9	9

a Sketch an appropriate graph to illustrate this data. **(3)**

b Explain what the graph you drew suggests about the data. **(2)**

c Identify the level of data used to measure the variables in the study. **(2)**

d Name a statistical test you could use to analyse the data in this study and explain why the test you have selected is appropriate. **(3)**

e If the result of the statistical analysis is significant at $p \leq 0.01$, explain whether the researcher should retain or reject the null hypothesis. **(3)**

7 A psychologist investigated how quickly people react in either noisy or silent conditions. In an experiment, participants sat in front of a screen on which 20 'five letter' combinations were displayed one at a time. Their task was to press the enter key each time the letter X appeared among the five letters (e.g. AHXKP requires the enter key to be pressed) or press the space bar if the letter X did not appear (e.g. YRTIF requires the space bar to be pressed).

Ten participants were in the silent condition and another ten participants were in the noisy condition. The noise was a recording of loud banging on drums. The reaction time was measured as how long it took for the participant to press the enter key or space bar and the computer calculated the average reaction time for each participant (in seconds). The findings are displayed in the bar chart.

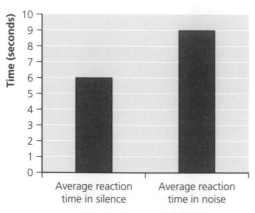

Figure E.1 Mean reaction time in silence and noise (seconds)

a State the mean reaction time taken in the silent condition. **(1)**

b Explain which group of participants had the fastest reaction time. **(2)**

c What level of data was gathered in this study? **(1)**

d Identify and simplify the ratio of the average reaction time in silence and the average reaction time in noise. **(2)**

e The psychologist analysed the data using the Mann–Whitney U test. With reference to the study, give two reasons for this choice of test. **(4)**

8 A psychologist investigated the effects of crowding on stress. A volunteer sample of 10 students were taken to a quiet country park and asked to walk around for 30 minutes. After their walk a sample of their saliva was taken. The following week the same students were taken to a very crowded city market and after 30 minutes in the crowd a sample of their saliva was taken. The cortisol level in the saliva samples was measured. (Cortisol is a stress hormone that is released when an individual interprets a situation as being stressful and can be measured in nanomoles per litre.) The results were as shown in Table E.7.

Table E.7 Level of cortisol in saliva after 30 minutes in uncrowded or crowded conditions

Participant	Cortisol level uncrowded (nmol/l)	Cortisol level crowded (nmol/l)
1	15	20
2	12	22
3	14	25
4	17	20
5	18	23
6	20	15
7	11	18
8	12	21
9	3	15
10	6	18
Mean	12.8	19.7
Range	17	10

a Calculate the median score for the uncrowded and crowded conditions. **(2)**

b Identify the modal score for the uncrowded and crowded conditions. **(2)**

c Suggest which measure of central tendency is the most appropriate for this set of data. **(3)**

d The range is given as a measure of dispersion. Identify one other measure of dispersion that could be used and outline one advantage of using this measure of dispersion. **(3)**

e Identify and justify which inferential statistical test should be used to analyse this data. **(4)**

f The psychologist proposed an experimental hypothesis: 'Levels of cortisol will be lower after 30 minutes in an uncrowded environment than after 30 minutes in a crowded environment'. Explain whether this is a one-tailed or two-tailed hypothesis. **(2)**

g If the statistical test finds the difference is significant at $p \leq 0.05$, explain whether the researcher should reject or retain the null hypothesis. **(4)**

9 A psychologist who was investigating the popularity of different types of ice-cream carried out an observational study watching people buying from an ice-cream van on a sunny Saturday afternoon. After watching for two hours, the data collected was as shown in Table E.8.

Table E.8

Type of ice-cream	Plain ice-cream cone	'99' ice-cream cone	Choc ice	Ice-lolly	Other
Number bought	35	40	15	50	25

a In total, how many ice-creams were bought? Show your working. **(2)**

b Which is the most popular type of ice-cream bought? **(1)**

c Explain which measure of central tendency is the most appropriate for this set of data. **(3)**

d What is the ratio of ice-cream cones sold to ice-lollies sold? Show your working. **(2)**

e What percentage of total sales were ice-lollies? Show your answer to two decimal places. **(2)**

10 A psychologist who was investigating fear of animals and insects handed out 280 questionnaires to an opportunity sample of sixth form students.

Question 1 was 'Are you afraid of one or more animal or insect?' YES/NO

Question 2 was 'Name one animal or insect you are afraid of.'

Question 3 was 'On a scale of 1 to 10, where 1 is "not very much" and 10 is "a great deal", how much do you fear the animal or insect that you named?'

Question 4 was 'Is any other member of your family afraid of the same animal or insect?' YES/NO

Some of the responses were as follows:

■ 32 respondents said they were afraid of dogs.

■ 25 respondents said they were afraid of spiders.

■ 15 respondents said they were afraid of wasps.

- 8 respondents said they were afraid of cats.
- 12 respondents said they were afraid of rats.
- 3 respondents said they were afraid of beetles.
- 45% of the participants identified at least one animal or insect that they feared.

a Calculate the percentage (to two decimal places) of participants who reported fearing

  i dogs

  ii spiders

  iii wasps

  iv cats   **(4)**

b If 45% of the participants identified at least one animal or insect that they feared, how many participants did NOT report fearing an animal or insect? **(2)**

c What is the ratio of participants who DID report fear to those who DID NOT report fear? **(2)**

Appendix 1

Cross referencing essential maths skills to research methods content

All GCE psychology specifications include the requirement for students to study research methods, scientific processes and techniques. Students must be familiar with the use of data handling and analysis techniques, be able to present data in graphs, tables, scatter diagrams and bar charts and be able to interpret these.

The cross reference gives suggestions as to how to relate the essential maths skills to the topics on the specification. These are only suggestions and many of the essential maths topics can be referred to, and relate to, other psychological content to be taught/learned. **The specification content is given in bold black.**

In the AS level cross reference, where a maths topic is indicated by * only OCR suggests this is assessed at AS level.

AS level specifications

Research methods

Experimental methods: types of experiment; laboratory and field experiments; natural and quasi experiments

Suggested maths skills:
- recognise and use expressions in decimal and standard form
- measures of central tendency – mean, median and mode
- calculate the arithmetic mean
- distinguish between levels of measurement *

Observational techniques: types of observation – naturalistic and controlled observation; covert and overt observation; participant and non-participant observation

Suggested maths skills:
- understand the differences between qualitative and quantitative data
- understand the difference between primary and secondary data
- distinguish between levels of measurement *

Self-report techniques: questionnaires; interviews – structured and unstructured

Suggested maths skills:
- understand the differences between qualitative and quantitative data
- understand the difference between primary and secondary data
- distinguish between levels of measurement *

Correlations: analysis of the relationship between co-variables; the difference between correlations and experiments

Suggested maths skills:
- use a scatter diagram to identify a correlation between two variables and plot two variables from experimental or other data
- select an appropriate statistical test *
- use a statistical test
- use statistical tables to determine significance *

Scientific processes

Aims and hypotheses: stating aims, the difference between aims and hypotheses; hypotheses – directional and non-directional

Suggested maths skills:
- use a statistical test
- use statistical tables to determine significance *

Sampling: the difference between population and sample; sampling techniques, including random, systematic, stratified, opportunity and volunteer; implications of sampling techniques, including bias and generalisation

Suggested maths skills:
- understand the principles of sampling as applied to scientific data
- use ratios, fractions and percentages

Pilot studies and the aims of piloting

Suggested maths skills:
- estimate results

Experimental designs: repeated measures, independent groups, matched pairs

Suggested maths skills:
- select an appropriate statistical test *
- use a statistical test
- use statistical tables to determine significance *

Observational design: behavioural categories; event sampling; time sampling

Suggested maths skills:
- construct and interpret frequency tables and diagrams, bar charts and histograms
- translate information between graphical, numeric and algebraic forms

Questionnaire construction, including use of open and closed questions; design of interviews

Suggested maths skills:
- understand the differences between qualitative and quantitative data
- understand the difference between primary and secondary data

Variables: manipulation and control of variables, including independent, dependent, extraneous, confounding; operationalisation of variables

Suggested maths skills:
- understand the differences between qualitative and quantitative data
- distinguish between levels of measurement *

Control: random allocation and counterbalancing, randomisation and standardisation. Demand characteristics and investigator effects

Suggested maths skills:
- understand the principles of sampling as applied to scientific data

Data handling and analysis

Quantitative and qualitative data: the distinction between qualitative and quantitative data collection techniques; primary and secondary data, including meta-analysis

Suggested maths skills:
- understand the differences between qualitative and quantitative data
- understand the difference between primary and secondary data

Descriptive statistics: measures of central tendency – mean, median, mode; calculation of mean, median and mode; measures of dispersion; range and standard deviation; calculation of range; calculation of percentages; positive, negative and zero correlations

Suggested maths skills:
- recognise and use expressions in decimal and standard form
- use ratios, fractions and percentages
- significant figures
- distinguish between levels of measurement *
- measures of central tendency – mean, median and mode
- calculate the arithmetic mean
- measures of dispersion including range and standard deviation

Presentation and display of quantitative data: graphs, tables, scatter diagrams, bar charts

Suggested maths skills:
- distinguish between levels of measurement *
- construct and interpret frequency tables and diagrams, bar charts and histograms
- translate information between graphical, numeric and algebraic forms
- use a scatter diagram to identify a correlation between two variables and plot two variables from experimental or other data

Distributions: normal and skewed distributions; characteristics of normal and skewed distributions

Suggested maths skills:
- measures of central tendency – mean, median and mode
- calculate the arithmetic mean
- measures of dispersion including range and standard deviation
- the characteristics of normal and skewed distributions
- calculate standard deviation
- understand and use the symbols: $=$, $<$, $\ll$, $\gg$, $>$, $\propto$, $\sim$
- substitute numerical values into algebraic equations *
- solve simple algebraic equations *

Introduction to statistical testing; the sign test

Suggested maths skills:

- recognise and use expressions in decimal and standard form
- understand and use the symbols: $=, <, \ll, \gg, >, \propto, \sim$
- substitute numerical values into algebraic equations *
- solve simple algebraic equations *
- understand simple probability
- measures of dispersion including range and standard deviation
- select an appropriate statistical test *
- use a statistical test
- use statistical tables to determine significance *

A-level specifications

Research methods

Experimental methods: types of experiment; laboratory and field experiments; natural and quasi experiments

Suggested maths skills:
- recognise and use expressions in decimal and standard form
- measures of central tendency – mean, median and mode
- calculate the arithmetic mean
- distinguish between levels of measurement

Observational techniques: types of observation – naturalistic and controlled observation; covert and overt observation; participant and non-participant observation

Suggested maths skills:
- understand the differences between qualitative and quantitative data
- understand the difference between primary and secondary data
- distinguish between levels of measurement

Self-report techniques: questionnaires; interviews – structured and unstructured

Suggested maths skills:
- understand the differences between qualitative and quantitative data
- understand the difference between primary and secondary data
- distinguish between levels of measurement

Correlations: analysis of the relationship between co-variables; the difference between correlations and experiments

Suggested maths skills:
- use a scatter diagram to identify a correlation between two variables and plot two variables from experimental or other data
- select an appropriate statistical test
- use a statistical test
- use statistical tables to determine significance

Content analysis and case studies

Suggested maths skills:

- understand the differences between qualitative and quantitative data
- understand the difference between primary and secondary data

Scientific processes

Aims and hypotheses: stating aims, the difference between aims and hypotheses; hypotheses – directional and non-directional

Suggested maths skills:

- use a statistical test
- use statistical tables to determine significance

Sampling: the difference between population and sample; sampling techniques, including random, systematic, stratified, opportunity and volunteer; implications of sampling techniques, including bias and generalisation

Suggested maths skills:

- understand the principles of sampling as applied to scientific data
- use ratios, fractions and percentages

Pilot studies and the aims of piloting

Suggested maths skills:

- estimate results

Experimental designs: repeated measures, independent groups, matched pairs

Suggested maths skills:

- select an appropriate statistical test
- use a statistical test
- use statistical tables to determine significance

Observational design: behavioural categories; event sampling; time sampling

Suggested maths skills:

- construct and interpret frequency tables and diagrams, bar charts and histograms
- translate information between graphical, numeric and algebraic forms

Questionnaire construction, including use of open and closed questions; design of interviews

Suggested maths skills:

- understand the differences between qualitative and quantitative data
- understand the difference between primary and secondary data

Variables: manipulation and control of variables, including independent, dependent, extraneous, confounding; operationalisation of variables

Suggested maths skills:

- understand the differences between qualitative and quantitative data
- distinguish between levels of measurement

Control: random allocation and counterbalancing, randomisation and standardisation. Demand characteristics and investigator effects

Suggested maths skills:
- understand the principles of sampling as applied to scientific data

Reliability across all methods of investigation: ways of assessing reliability – test-retest and inter-observer; improving reliability

Suggested maths skills:
- understand the differences between qualitative and quantitative data
- understand the difference between primary and secondary data

Types of validity across all methods of investigation: face validity, concurrent validity, ecological validity and temporal validity, assessment of validity, improving validity

Suggested maths skills:
- understand the differences between qualitative and quantitative data
- understand the difference between primary and secondary data
- understand the principles of sampling as applied to scientific data
- distinguish between levels of measurement

Features of science: objectivity and the empirical method; replicability and falsifiability; theory construction and hypothesis testing; paradigms and paradigm shifts

Suggested maths skills:
- understand the differences between qualitative and quantitative data
- understand the difference between primary and secondary data
- understand simple probability
- measures of dispersion including range and standard deviation
- the characteristics of normal and skewed distributions
- calculate standard deviation
- select an appropriate statistical test
- use a statistical test
- use statistical tables to determine significance

Data handling and analysis

Quantitative and qualitative data: the distinction between qualitative and quantitative data collection techniques; primary and secondary data, including meta-analysis

Suggested maths skills:
- understand the differences between qualitative and quantitative data
- understand the difference between primary and secondary data

Descriptive statistics: measures of central tendency – mean, median, mode; calculation of mean, median and mode; measures of dispersion; range and standard deviation; calculation of range; calculation of percentages; positive, negative and zero correlations

Suggested maths skills:

- recognise and use expressions in decimal and standard form
- use ratios, fractions and percentages
- significant figures
- distinguish between levels of measurement
- measures of central tendency – mean, median and mode
- calculate the arithmetic mean
- measures of dispersion including range and standard deviation

Presentation and display of quantitative data: graphs, tables, scatter diagrams, bar charts and histograms

Suggested maths skills:

- distinguish between levels of measurement
- construct and interpret frequency tables and diagrams, bar charts and histograms
- translate information between graphical, numeric and algebraic forms
- use a scatter diagram to identify a correlation between two variables and plot two variables from experimental or other data

Analysis and interpretation of correlation, including correlation coefficients

Suggested maths skills:

- use a scatter diagram to identify a correlation between two variables and plot two variables from experimental or other data
- select an appropriate statistical test
- use a statistical test
- use statistical tables to determine significance

Levels of measurement: nominal, ordinal and interval

Suggested maths skills:

- distinguish between levels of measurement
- measures of central tendency – mean, median and mode
- calculate the arithmetic mean and/or median and/or identify modal scores

Distributions: normal and skewed distributions; characteristics of normal and skewed distributions

Suggested maths skills:

- measures of central tendency – mean, median and mode
- calculate the arithmetic mean
- measures of dispersion including range and standard deviation
- the characteristics of normal and skewed distributions
- calculate standard deviation
- understand and use the symbols: $=$, $<$, $\ll$, $\gg$, $>$, $\propto$, $\sim$
- substitute numerical values into algebraic equations
- solve simple algebraic equations

Inferential testing

Students should demonstrate knowledge and understanding of inferential testing and be familiar with the use of inferential tests.

Introduction to statistical testing; the sign test

Suggested maths skills:
- recognise and use expressions in decimal and standard form
- understand and use the symbols: $=, <, \ll, \gg, >, \propto, \sim$
- substitute numerical values into algebraic equations
- solve simple algebraic equations
- understand simple probability
- use a statistical test
- use statistical tables to determine significance

Probability and significance: use of statistical tables and critical values in interpretation of significance; Type 1 and Type 2 errors

Suggested maths skills:
- recognise and use expressions in decimal and standard form
- understand and use the symbols: $=, <, \ll, \gg, >, \propto, \sim$
- substitute numerical values into algebraic equations
- solve simple algebraic equations
- understand simple probability
- measures of dispersion including range and standard deviation
- select an appropriate statistical test
- use a statistical test
- use statistical tables to determine significance

Factors affecting the choice of statistical test, including level of measurement and experimental design; when to use the following tests – Spearman's Rho, Pearson's R, Wilcoxon, Mann–Whitney, Related T test, Unrelated T test and Chi-square test

Suggested maths skills:
- distinguish between levels of measurement
- the characteristics of normal and skewed distributions
- calculate standard deviation
- select an appropriate statistical test
- use a statistical test
- use statistical tables to determine significance

Appendix 2

Formulae and statistical tables

Standard deviation (sample)

$$s = \sqrt{\frac{\sum(x - \bar{x})^2}{n - 1}}$$

Spearman's Rho (Rank Order) correlation coefficient

$$r = 1 - \frac{6\sum d^2}{n(n^2 - 1)}$$

Critical values for Spearman's Rho

	Level of significance for a one-tailed test				
	0.05	0.025	0.01	0.005	0.0025
	Level of significance for a two-tailed test				
n	0.10	0.05	0.025	0.01	0.005
4	1.000	1.000	1.000	1.000	1.000
5	0.700	0.900	0.900	1.000	1.000
6	0.657	0.771	0.829	0.943	0.943
7	0.571	0.679	0.786	0.857	0.893
8	0.548	0.643	0.738	0.810	0.857
9	0.483	0.600	0.683	0.767	0.817
10	0.442	0.564	0.649	0.733	0.782
11	0.418	0.527	0.609	0.700	0.755
12	0.399	0.504	0.587	0.671	0.727
13	0.379	0.478	0.560	0.648	0.698
14	0.367	0.459	0.539	0.622	0.675
15	0.350	0.443	0.518	0.600	0.654
16	0.338	0.427	0.503	0.582	0.632
17	0.327	0.412	0.482	0.558	0.606
18	0.317	0.400	0.468	0.543	0.590
19	0.308	0.389	0.456	0.529	0.575
20	0.299	0.378	0.444	0.516	0.561
21	0.291	0.369	0.433	0.503	0.549
22	0.284	0.360	0.423	0.492	0.537
23	0.277	0.352	0.413	0.482	0.526
24	0.271	0.344	0.404	0.472	0.515
25	0.265	0.337	0.396	0.462	0.505
26	0.260	0.330	0.388	0.453	0.496
27	0.255	0.323	0.381	0.445	0.487
28	0.250	0.317	0.374	0.437	0.479
29	0.245	0.312	0.367	0.430	0.471
30	0.241	0.306	0.361	0.423	0.463

The calculated value must be equal to or exceed the critical value in this table for significance to be shown.

Chi-squared distribution formula

$$x^2 = \sum \frac{(o-e)^2}{e}$$

$$df = (r-1)(c-1)$$

Critical values for Chi-squared distribution

	Level of significance for a one-tailed test					
	0.10	0.05	0.025	0.01	0.005	0.0005
	Level of significance for a two-tailed test					
df	0.20	0.10	0.05	0.025	0.01	0.001
1	1.64	2.71	3.84	5.02	6.64	10.83
2	3.22	4.61	5.99	7.38	9.21	13.82
3	4.64	6.25	7.82	9.35	11.35	16.27
4	5.99	7.78	9.49	11.14	13.28	18.47
5	7.29	9.24	11.07	12.83	15.09	20.52
6	8.56	10.65	12.59	14.45	16.81	22.46
7	9.80	12.02	14.07	16.01	18.48	24.32
8	11.03	13.36	15.51	17.54	20.09	26.12
9	12.24	14.68	16.92	19.02	21.67	27.88
10	13.44	15.99	18.31	20.48	23.21	29.59
11	14.63	17.28	19.68	21.92	24.73	31.26
12	15.81	18.55	21.03	23.34	26.22	32.91
13	16.99	19.81	22.36	24.74	27.69	34.53
14	18.15	21.06	23.69	26.12	29.14	36.12
15	19.31	22.31	25.00	27.49	30.58	37.70
16	20.47	23.54	26.30	28.85	32.00	39.25
17	21.62	24.77	27.59	30.19	33.41	40.79
18	22.76	25.99	28.87	31.53	34.81	42.31
19	23.90	27.20	30.14	32.85	36.19	43.82
20	25.04	28.41	31.41	34.17	37.57	45.32
21	26.17	29.62	32.67	35.48	38.93	46.80
22	27.30	30.81	33.92	36.78	40.29	48.27
23	28.43	32.01	35.17	38.08	41.64	49.73
24	29.55	33.20	36.42	39.36	42.98	51.18
25	30.68	34.38	37.65	40.65	44.31	52.62
26	31.80	35.56	38.89	41.92	45.64	54.05
27	32.91	36.74	40.11	43.20	46.96	55.48
28	34.03	37.92	41.34	44.46	48.28	56.89
29	35.14	39.09	42.56	45.72	49.59	58.30
30	36.25	40.26	43.77	46.98	50.89	59.70
40	47.27	51.81	55.76	59.34	63.69	73.40
50	58.16	63.17	67.51	71.42	76.15	86.66
60	68.97	74.40	79.08	83.30	88.38	99.61
70	79.72	85.53	90.53	95.02	100.43	112.32

The calculated value must be equal to or less than the critical value in this table for significance to be shown.

Mann–Whitney U test formulae

$$U_a = n_a n_b + \frac{n_a(n_a + 1)}{2} - \sum R_a$$

$$U_b = n_a n_b + \frac{n_b(n_b + 1)}{2} - \sum R_b$$

(U is the smaller of U_a and U_b)

Critical values for the Mann–Whitney U test

	N_b															
	5	6	7	8	9	10	11	12	13	14	15	16	17	18	19	20
N_a																
P ≤ 0.05 (one-tailed), p ≤ 0.10 (two-tailed)																
5	4	5	6	8	9	11	12	13	15	16	18	19	20	22	23	25
6	5	7	8	10	12	14	16	17	19	21	23	25	26	28	30	32
7	6	8	11	13	15	17	19	21	24	26	28	30	33	35	37	39
8	8	10	13	15	18	20	23	26	28	31	33	36	39	41	44	47
9	9	12	15	18	21	24	27	30	33	36	39	42	45	48	51	54
10	11	14	17	20	24	27	31	34	37	41	44	48	51	55	58	62
11	12	16	19	23	27	31	34	38	42	46	50	54	57	61	65	69
12	13	17	21	26	30	34	38	42	47	51	55	60	64	68	72	77
13	15	19	24	28	33	37	42	47	51	56	61	65	70	75	82	84
14	16	21	26	31	36	41	46	51	56	61	66	71	77	82	87	92
15	18	23	28	33	39	44	50	55	61	66	72	77	83	88	94	100
16	19	25	30	36	42	48	54	60	65	71	77	83	89	95	101	107
17	20	26	33	39	45	51	57	64	70	77	83	89	96	102	109	115
18	22	28	35	41	48	55	61	68	75	82	88	95	102	109	116	123
19	23	30	37	44	51	58	65	72	80	87	94	101	109	116	123	130
20	25	32	39	47	54	62	69	77	84	92	100	107	115	123	130	138

	N_b															
	5	6	7	8	9	10	11	12	13	14	15	16	17	18	19	20
N_a																
P ≤ 0.01 (one-tailed), p ≤ 0.02 (two-tailed)																
5	1	2	3	4	5	6	7	8	9	10	11	12	13	14	15	16
6	2	3	4	6	7	8	9	11	12	13	15	16	18	19	20	22
7	3	4	6	7	9	11	12	14	16	17	19	21	23	24	26	28
8	4	6	7	9	11	13	15	17	20	22	24	26	28	30	32	34
9	5	7	9	11	14	16	18	21	23	26	28	31	33	36	38	40
10	6	8	11	13	16	19	22	24	27	30	33	36	38	41	44	47
11	7	9	12	15	18	22	25	28	31	34	37	41	44	47	50	53
12	8	11	14	17	21	24	28	31	35	38	42	46	49	53	56	60
13	9	12	16	20	23	27	31	35	39	43	47	51	55	59	63	67
14	10	13	17	22	26	30	34	38	43	47	51	56	60	65	69	73
15	11	15	19	24	28	33	37	42	47	51	56	61	66	70	75	80
16	12	16	21	26	31	36	41	46	51	56	61	66	71	76	82	87
17	13	18	23	28	33	38	44	49	55	60	66	71	77	82	88	93
18	14	19	24	30	36	41	47	53	59	65	70	76	82	88	94	100
19	15	20	26	32	38	44	50	56	63	69	75	82	88	94	101	107
20	16	22	28	34	40	47	53	60	67	73	80	87	93	100	107	114

N_b	5	6	7	8	9	10	11	12	13	14	15	16	17	18	19	20
N_a																
$P \leq 0.025$ (one-tailed), $p \leq 0.05$ (two-tailed)																
5	2	3	5	6	7	8	9	11	12	13	14	15	17	18	19	20
6		5	6	8	10	11	13	14	16	17	19	21	22	24	25	27
7			8	10	12	14	16	18	20	22	24	26	28	30	32	34
8				13	15	17	19	22	24	26	29	31	34	36	38	41
9					17	20	23	26	28	31	34	37	39	42	45	48
10						23	26	29	33	36	39	42	45	48	52	55
11							30	33	37	40	44	47	51	55	58	62
12								37	41	45	49	53	57	61	65	69
13									45	50	54	59	63	67	72	76
14										55	59	64	67	74	78	83
15											64	70	75	80	85	90
16												75	81	86	92	98
17													87	93	99	105
18														99	106	112
19															113	119
20																127

N_b	5	6	7	8	9	10	11	12	13	14	15	16	17	18	19	20
N_a																
$P \leq 0.005$ (one-tailed), $p \leq 0.01$ (two-tailed)																
5	0	1	1	2	3	4	5	6	7	7	8	9	10	11	12	13
6		2	3	4	5	6	7	9	10	11	12	13	15	16	17	18
7			4	6	7	9	10	12	13	15	16	18	19	21	22	24
8				7	9	11	13	15	17	18	20	22	24	26	28	30
9					11	13	16	18	20	22	24	27	29	31	33	36
10						16	18	21	24	26	29	31	34	37	39	42
11							21	24	27	30	33	36	39	42	45	48
12								27	31	34	37	41	44	47	51	54
13									34	38	42	45	49	53	57	60
14										42	46	50	54	58	63	67
15											51	55	60	64	69	73
16												60	65	70	74	79
17													70	75	81	86
18														81	87	92
19															93	99
20																105

The calculated value must be equal to or less than the critical value in this table for significance to be shown.

Wilcoxon matched pairs signed ranks test process

- Calculate the difference between each pair of scores by subtracting one from the other.
- Rank the differences:
  - The smallest difference is Rank 1.
  - Ignore any zero differences.
  - Ignore the sign (positive or negative) of the difference.
- Add together the ranks with a positive sign.
- Add together the ranks with a negative sign.
- T is the smaller of these two totals. It may be positive or negative.
- N is the number of differences (excluding zero differences).

Critical values for the Wilcoxon matched pairs signed ranks test

	Level of significance for a one-tailed test		
	0.05	0.025	0.01
	Level of significance for a two-tailed test		
n	0.1	0.05	0.02
N=5	0	-	-
6	2	0	-
7	3	2	0
8	5	3	1
9	8	5	3
10	11	8	5
11	13	10	7
12	17	13	9

The calculated value must be equal to or less than the critical value in this table for significance to be shown.